Improving Maintenance & Reliability Through Cultural Change

Stephen J. Thomas

Industrial Press

Library of Congress Cataloging-in-Publication Data
Thomas, Stephen J.
    Improving maintenance and reliability through cultural change / Stephen J.
Thomas.
        p. cm.
    Includes bibliographical references and index.
    ISBN 0-8311-3190-X
    1. Reliability (Engineering). 2. Maintainability (Engineering). 3.
Organizational change--Management. I. Title.
    TA169.T52 2005
    658.2'02--dc
                                                                2005047412

Industrial Press Inc.

989 Avenue of the Americas, New York, NY 10018
Tel: 212-889-6330 Toll-Free: 1-888-528-7852 Fax: 212-545-8327
www.industrialpress.com Email: info@industrialpress.com

First Edition, April 2005

Sponsoring Editor: John Carleo
Interior Text and Cover Design: Janet Romano
Developmental Editor: Robert Weinstein

This book is dedicated to
Jacob Elan Thomas, my grandson,
born November 1, 2004.

Table of Contents

Foreword

Improving Maintenance and Reliability Through Cultural Change

Many years ago I began a career in engineering at a nuclear power generation utility company. My duties very soon focused on plant maintenance and reliability, and I have been deeply involved in industrial plant maintenance and reliability ever since. Eighteen years ago I founded Management Resources Group, Inc. MRG is a maintenance and reliability professional services company, which now boasts nearly half the Fortune 500 as clients. Since founding MRG, my travels have taken me to many corners of the world in more than a dozen different industries, observing and helping improve plant maintenance and reliability practices at over 300 plants. I have learned a lot in that time, and I still learn something new just about every day. But one of my most important lessons relates to the subject of this book – organizational culture.

Although the science and technology we use to accomplish maintenance and to optimize equipment reliability have been honed over the years, one fact remains the same, and is the source of frustration wherever I go. That is, no matter how rational and sensible the business case for embracing different (and presumably better) practices may be, getting the people to adopt the practices and adapt to a new and better way of doing business, is always extremely difficult. I refer to this difficulty as the "softer side" of the problem. Until the author, Stephen J. Thomas, put some handles on this "softer side" for me with his first book, *"Successfully Managing Change in Organizations, A User's Guide,"* I was under the wrong impression that scientific tools and methodologies applicable to the "soft side" did not exist. Mr. Thomas showed us in his first book that there are indeed such tools. They come to us from the adult learning, psychology, behavioral sciences and related disciplines. I now know that overlaying these tools and methods atop any major change initiative – especially one related to maintenance and reliability – will enhance an organization's ability to achieve a new way of doing business – to change the habits of people, and in essence, make new practices stick.

xi

If you want a vivid example of organizational culture that you can identify with, let's talk about "safety." When I first came into the workforce, "safety" was considered a "silo" of the organization – with particular people given responsibility to ensure that safety was taken seriously, and that injuries were eliminated. Back then, "safety" was not everyone's job – it was a responsibility of a limited number of employees in the company who were trying to get people to work smarter and safer. It took many, many years for today's leading industrial companies to arrive at a point where "safety" is indeed *everyone's* job and *everyone's* responsibility. Today, safety is no longer the domain of a silo or department in the organization. Today, "safety" is embedded into the fabric of the company – it is how we do business – it doesn't require constant pressure to sustain – it is a part of the organization's Culture!

Think about it – what did we do in these companies to make safety such an integrated part of the company's culture? What did we do to make that change stick? How did it become the new way of doing business – the new status quo – now taken for granted? How difficult would it be to try to change the safety culture of your organization today? Pretty tough, I suspect.

I believe the way we do maintenance and reliability can, and will have to, become a part of the fabric and culture of our companies if we are to succeed in our personal goals, and if our companies are to succeed in their business goals.

How do we change the maintenance and reliability practices of our organizations? The change management tools presented in Mr. Thomas's first book provides ample guidance on that question. However, changing practices and habits of the people also requires understanding and dealing with (and possibly changing) the organizational culture, and *organizational culture* is the subject of this book. The fact is the best engineering, science, and technology won't ensure business success in maintenance and reliability. Without these "hard" tools, we won't have success – granted. But these "hard" tools alone won't ensure success. The *culture* of your organization has to be understood and dealt with.

In his newest book, Mr. Thomas has really captured and organized for our use the definitions, tools, methods, elements, case studies, practical guides, and templates that will help us understand organizational culture. Building on the eight elements of the Web of Change from the last book, this new book deals directly with first understanding an organization's culture, then measuring its readiness for change, and finally sustaining change until it becomes the new culture of the organization.

I met the author, Stephen J. Thomas, many years ago during a major, company-wide maintenance and reliability change initiative that he was then leading. We soon became close associates and eventually enduring friends. I respect Steve immensely and consider him to be one of the smartest, most professional and most trustworthy people I know. Steve's accomplishments in his long career are remarkable, and I am very proud of him and what he has accomplished — and continues to accomplish today. I am most proud of his generous ethic that drives him to share his experiences with the rest of us, in the interest of helping others succeed in improving business performance through culture change.

I am very enthusiastic about this book because I know it will help many people change the culture of their organizations. These changes will enable improved maintenance and reliability practices to take hold, helping industrial companies reduce the cost and improve the quality of their products – and this will ultimately improve the business performance of those companies. I have had the benefit of reading this book and I can assure you that it will help you deal with the "softer side" of changing the way your company does maintenance and reliability business. I hope you enjoy it and put it to practical use – I have!

Congratulations, Steve, on another important, practical, and valuable contribution to this maintenance and reliability profession of ours!

> Robert S. DiStefano, CMRP
> Chairman of the Board
> Management Resources Group, Inc.

Acknowledgements

I would like to express my sincere thanks to my publisher Industrial Press, Inc. specifically John Carleo for your faith in my work and the potential value it can deliver, and to Janet Romano for all of the behind-the-scenes work that has gone into the excellent finished product.

I would also like to thank Robert Weinstein for helping provide clarity and focus in order to deliver a better product.

The Beginning

1.1 Introduction

One of the major areas of focus in industry today is that of improving equipment reliability. Why? To insure that production is always available to meet the demand of the marketplace. One of the worst nightmares of any company and those who manage it is to have a demand for product but not be able to supply it because of equipment failure. Certainly this scenario will reduce company profitability and could ultimately put a company out of business.

For some firms, poor reliability and its impact on production are far more serious than for others. For those that operate on a continuous basis – they run 24 hours per day seven days per week – there is no room for unplanned shutdowns of the production equipment; any loss of production is often difficult or even impossible to make up. For others that do not operate in a 24/7 mode, recovery can be easier, but nevertheless time consuming and expensive, reducing profits.

Many programs available in the industry are designed to help businesses improve reliability. They are identified in trade literature, promoted at conferences and over the web, and quite often they are in place within the plants in your own company. Most of these programs are what I refer to as "hard skill" programs. They deal with the application of resources and resource skills in the performance of a specific task aimed at reliability improvement. For example, you decide that you want to improve preventive maintenance (PM). To accomplish this you train your workforce in preventive maintenance skills, purchase the necessary equipment, and roll out a PM program accompanied with corporate publicity, presentations of what you expect to accomplish, and other forms of hype in order to get buy in from those who need to execute it. Then you congratulate your team for a job well done and move on to the next project.

Often at this juncture, something very significant happens. The program you delivered starts strongly, but immediately things begin to go wrong. The work crews assigned to preventive maintenance get diverted to other plant priorities; although promises are made to return them to their original PM assignments, this never seems to happen. Equipment that is scheduled to be out of service for preventive maintenance can't be shutdown due to the requirements of the production department; although promises are made to take the equipment off-line at a later time, this never seems to happen. Finally, the various key members of management who were active advocates and supporters at the outset are the very ones who permit the program interruptions, diminish its intent, and reduce the potential value. Often these people do make attempts to get the program back on track, but these attempts are often half-hearted. Although nothing is openly said, the organization recognizes what is important, and often this is not the preventive maintenance program.

I have simplified the demise of the preventive maintenance program in our example. Yet this is exactly as it happens, although much more subtle. In the end, the result is the same. Six months after the triumphant rollout of the program, it is gone. The operational status quo has returned and, if you look at the business process, you may not even be able to ascertain that a preventive maintenance program ever existed at all.

For those of us trying to improve reliability or implement any type of change in our business, the question we need to ask ourselves is why does this happen? The intent of the program was sound. It was developed with a great deal of detail, time, and often money; the work plan was well executed. Yet in the end there is nothing to show for all of the work and effort.

Part of the answer is that change is a difficult **process.** Note that I didn't say **program**, because a program is something with a beginning and an end. A process has a starting point – when you initially conceived the idea – but it has no specific ending and can go on forever.

Yet the difficulty of implementing change isn't the root cause of the problem. You can force change. If you monitor and take proper corrective action, you may even be able over the short term to force the process to appear successful. Here, the operative word is you. What if you implement the previously-mentioned preventive maintenance program and then, in order to assure compliance, continually monitor the progress. Further suppose that you are a senior manager and have the ability to rapidly remove from the process change any roadblocks it encounters as it progresses. What then? Most likely the change will stick as long as you are providing care and feeding. But what do you think will happen if after one month into the program you are removed from the equation. If there are no other supporters to continue the oversight and corrective action efforts, the program will most likely lose energy. Over a relatively short time, everything will likely return to the status quo.

The question we need to answer is why does this happen to well-intentioned reliability-driven change process throughout industry? The answer is that the process change is a victim of the organization's culture. This hidden force, which defines how an organization behaves, works behind the scenes to restore the status quo unless specific actions are taken to establish a new status quo for the organization. Without proper attention to organizational culture, long-term successful change is not possible.

Think of an organization's culture as a rubber band. The more you try to stretch it, the harder it tries to return to its previously unstretched state – the status quo. However if you stretch it in a way that it can't return and leave it stretched for a long time, once released it will retain the current stretch you gave it and not return to its original dimensions. That is going to be our goal – to figure out how to stretch the organization, but in a way that when the driver of the process is out of the equation, the organizational rubber band won't snap back.

1.2 Why I Wrote This Book

I have been involved in the world of change management for sometime. Quite often I have been involved in projects that were well planned and executed and then, when attention was turned to other initiatives, they disappeared in the blink of an eye. I as well as a great many others had difficulty understanding why this happened. Were the projects or effort ill conceived? Not really. They addressed specific business needs and had the potential for delivering value to the company. Were they poorly designed and not rolled out to the organization properly? No, some of the smartest people I know were involved in many of the efforts. Was the organization out to destroy the effort? Actually when these projects were rolled out, there usually was a lot of enthusiasm and interest from the organization. So what was the cause of the failure?

Some of the work I did on my last book, *Successfully Managing Change in Organizations: A Users Guide*, explored the causes by examining the **eight elements of change** – leadership, work process, structure, group learning, technology, communication, interrelationships, and rewards in a holistic fashion via **The Web of Change**. From this and other work I have done, I recognized that properly addressing the eight elements was very important. But beyond these there are two concepts that are critical success factors to the longevity of any change. These are readiness and sustainability.

Coupled with my book and workbook was a training program using my concepts to help deliver successful change. However, as I developed training plans, I recognized that in training people to use new ideas and concepts, readiness and sustainability were two important drivers for success. However, knowing this raises a question. Why are readiness and sustainability required to help implement newly-learned ideas in a plant or any other setting?

It was around this time that I had several discussions with other change practitioners. Although they were not talking about readiness or sustainability, it was clear that both of these elements were an important part of a much larger concept. This concept focused on the need to change the organization's culture in order to successfully implement new initiatives.

This raised questions in my mind. What is organizational culture and why do you need to change it to have a successful change process implementation? To answer this question, I began doing research on the topic of cultural change. What I discovered is that the concept is talked about, but what it really is and how you go about changing it was not crystal clear. In my search, I again ran into three types of books on this subject, as I had found during my research on my previous book. These books included CEOs describing how they made change at the senior management level; books by academics who explained the concept, but in ways not applicable to middle management; and consultants with many good ideas, some of which were available in the book, but others that required a fee for services.

The conclusion – if you really want to have successful change, changing organization culture is critical, but practical information for those of us in middle management who have to make it happen is lacking.

As a result of my investigation, I found the same motivation for this book as I did for my last book. What I have tried to create in the chapters that follow is a book designed for those of us in middle management who are faced with the day-to-day task of implementing successful change; we need to understand how to change culture so that the initiatives that we are being asked to implement are not only successful at the outset, but provide long-lasting change for the business.

The goal of this book is to show you how to change the culture in order to promote readiness, sustainability, and ultimate success for whatever initiative you are rolling out to the organization.

1.3 Who Can Benefit

Changing the organization's culture in order to promote long-lasting change benefits everyone from the top of the organization to the bottom. It benefits the top by providing a solid foundation on which to build new concepts, behaviors, and ways of thinking about work. To accomplish radical changes in how we think about or execute our work requires that those who are part of the culture support it. Senior management can only take this so far. They can set and communicate the vision and they can visibly support the effort, but the most important thing they can do is empower those in the middle to make it happen.

As middle managers, we all know that we have many more initiatives on which to work than there is time in the day (or night). This spreads our focus. If we don't collectively embrace the new change, then no threat, benefit or any other motivational technique will make the change successful over the long term. Although this book can help educate senior management so that they can empower the rest of us, its real benefit is for middle managers. It will help them understand this very complex concept in a way that will enable them to deliver successful change initiatives.

Just as with *Successfully Managing Change in Organizations: A Users Guide*, I have been unable to find a book that I believe adequately explains the concept of organizational culture to those of us in the middle who are asked to change it. This book provides middle managers with "how to" guidance for cultural change," targeting reliability improvement.

There is also benefit for those at the **bottom** tier of the management hierarchy. The term bottom is not meant to demean this roll because this is where the "rubber meets the road." All cultural change and their related initiatives end here. This is where all of the plans, training, and actual work to implement end. If it doesn't work here – failure is the outcome. The benefit of this book for you is that this book, unlike any other out there on the market, explains the concepts of cultural change in a manner that is clearly understandable and applicable to the difficult task you have in front of you every day.

1.4 Why Is This Book Different?

When I first heard the expression **organizational culture,** I had no idea what it meant. What I later learned was that the people telling me that we had to "change the organization's culture" were not completely sure either. They sensed that, in addition to the skills we were trying to impart and the processes we are trying to change, there was something more basic that, if not addressed, had the potential to impede or even totally destroy what we were trying to accomplish.

The problem was that we were not entirely clear on what that something actually was, yet we knew it was there. As I did more research and questioning of others, I became convinced that people are very much aware of this key ingredient, yet most are not sure how to define it and get their arms around it. There is a general lack of clarity about the concept of cultural change, what is involved to successfully accomplish it, and what parts need to be addressed if one is to be successful.

This book will demystify the concept of cultural change. It will explain organizational culture in a way that will make the concept clear to everyone. As a result, it will enable people to know what to ask for, what to do, and how to implement changes to the very fabric of the organization for sustainable improvement over the long haul.

The book is unique not only in that it makes a complex process easily understood, but it also brings into the equation the **eight elements of change** and shows how they relate to and support the concept.

To add one other benefit, this book is focused on a specific aspect of change in the work place, namely equipment reliability. There are many books on this subject, but none that tie the concept of cultural change to that of improving reliability of plant equipment.

1.5 What Is Included?

If you have read my book *Successfully Managing Change in Organizations: A Users Guide,* you will recall that there are many topics that I addressed in detail such as vision, the Goal Achievement Model, outcomes and impacts, and a methodology for portraying the health of your change process through **The Web of Change**. All of these topics are pertinent to the study of cultural change; although I will reference them, I will refer you to my previous book for the details.

Also in *Successfully Managing Change in Organizations: A Users Guide,* I discussed the **eight elements of change**. Each of these – leadership, work process, structure, group learning, technology, communication, interrelationships, and rewards – is important in its own right, but collectively they are even more important in how they relate to one another. In my prior book, all eight elements were addressed in two chapters. This was the short version, but it served to help the reader understand the concept and the connectivity between the elements. In this book, as you will see in the chapter synopsis below, I have addressed each of the elements in its own chapter and will tie them closely to the theme of this book – organization culture and cultural change.

I will also provide you with a tool, **The Cultural Web of Change,** to assess the progress you are making towards changing the culture of your organization. In this way you will be able to identify the gaps. Then using Change Root Cause Failure Analysis (C-RCFA), you will be able to take corrective action to get your process back on track. This tool will be described in Chapter 17 and is included on the CD at the back of the book.

In addition, the CD contains a file (with voice) providing the reader with an oversight of the topic and material contained in the text. This presentation will help you and those within your company get an initial understanding of the content.

The chapters that follow should prove to be interesting and enlightening. They will include real examples of my own and those of key individuals with whom I discussed this topic so you can get a broader perspective on the issues.

Chapter 1 Getting Started

This chapter opens up the topic of organizational culture as it relates to plant reliability. In *Successfully Managing Change in Organizations: A Users Guide.* I discussed the **eight elements of change**. This book takes the next step and introduces the glue that binds the eight elements together – culture.

Chapter 2 Culture Defined

The first step towards understanding culture is to define it. This chapter does just that and introduces the four elements of culture – values, role models, rites and rituals, and cultural infrastructure. These will each be addressed separately in Chapters 4 through 7.

Chapter 3 The Need for Vision and Goals

Equally important in the study of culture are the concepts of vision and goals. The vision sets the overall strategic direction that the firm intends to pursue over the long haul. The goals, or in our case the Goal Achievement Model, is how you get from where you are to where you want to be. These two concepts are foundational and are what an organization's cultural change process is built upon.

Chapter 4 Organizational Values

Organizational values are how a company communicates to its employees what is important. It enables the workforce to be able to make the right decisions when faced with a number of choices. This chapter provides clarity to this topic.

Chapter 5 Role Models

In the area of work performance, what gets measured gets done. The same maxim applies to human behavior. In this case, how the boss or senior management behaves gets modeled or replicated by those in the organization. Role models are a critical aspect of culture. In changing a culture, specific behaviors need to be instilled in the organization. Because we know that people model behavior of the leadership, these individuals play a crucial role in promulgating the new and specific cultural behaviors.

Chapter 6 Rites and Rituals

Rites and rituals are the third major component of organizational culture. They are two separate concepts but ones that reinforce each other. As defined by Deal and Kennedy in their landmark text **Corporate Cultures,** rituals are how things are done, not just the major things, but all things. Rituals communicate to everyone the accepted behavior within the context of a company culture. Rituals can even exist at the sub-culture level, but they provide the same level of employee behavioral guidance. Understanding rituals is the key to understanding how to change them.

Rites, the second component, are how the company reinforces acceptable behavior. A grasp of both of these and how they support one another is a good first step towards cultural change.

Chapter 7 Cultural Infrastructure

This chapter introduces a very important concept in the review of organizational culture. The cultural infrastructure is an invisible set of processes that help to shape organizational behavior and communication. If you don't understand and pay attention to these forces, you will discover that changes you want to institute take a lot longer than expected and some, in spite of all of your efforts, do not get done at all

Chapter 8 Introduction to the Elements

This chapter bridges the discussion in my last book, *Successfully Managing Change in Organizations: A Users Guide* and the more detailed and culturally focused discussion of these elements in this book. The eight elements are the same, but this book will explore each in more depth and focus on how they individually and collectively support cultural change.

Chapter 9 Leadership

Leadership is the most important of the eight elements. Chapter 9 is written in two major parts. Part 1 provides you with a broad understanding of the term leadership and how it is applied. Part 2 focuses on leadership as a critical part of the eight elements of change and how it impacts the four elements of culture.

Chapter 10 - Work Process

Work process has actually has three parts – information flow, the actual flow of work, and the flow of materials in a company. The first of these three components will be addressed in the chapter on communication. The third will not be a topic in this text because it ties more closely to a text on logistics. However the second component is key to cultural change. The reason for this is that, in many ways, the processes associated with doing work differently are exactly what cultural change is all about. This chapter will discuss ways to change culture by changing the work process.

Chapter 11 - Structure

How organizations are structured is often the difference between success and failure. Structure defines in many ways how things get accomplished. Reporting relationships and how functions are tied together are very important in changing a culture. This chapter looks at structure with a cultural focus and shows how to support change by the proper structural alignment.

Chapter 12 - Group Learning

One aspect of successful cultural change is tied to how we develop goals and initiatives, how we work to accomplish them, and most importantly how we learn from our outcomes and then re-apply what we learn to enhance our future work efforts. Nothing in the world of change happens linearly. We execute, review the outcome, and then make adjustments. This chapter will explore both single and double loop learning as it relates to cultural change.

Chapter 13 - Technology

In our case, technology applies to software applications that support work processes. In the maintenance and reliability arena, these applications usually are the computerized maintenance management systems and others that support the process. Technology in this context is an integral part of the cultural infrastructure by supporting or not supporting how things get done. Technology also plays an important supporting part in the organization's business rituals.

Chapter 14 – Communication

This chapter addresses how information flows throughout the organization. This is important because work can not be accomplished successfully without proper communication. Neither can any change in organizational culture. Communication is the heart of the matter; without it, a culture can not be altered.

Chapter 15 - Interrelationships

Like communications, interrelationships help or hinder how things are accomplished in business. This element touches all four parts of the cultural model. Our discussion in this chapter will deal with understanding interrelationships and being able to apply this understanding to changing culture.

Chapter 16 - Rewards

Everyone knows that what gets measured gets done. How we reward what gets done is equally important. If we measure to get people to pay attention to specific outcomes and then reward or reinforce appropriately, we will have increased the likelihood of these new behaviors continuing. Conversely, inappropriate rewards can have the exact opposite effect.

Chapter 17 The Web of Cultural Change

This chapter describes and explains the use of the cultural web model. This web diagram has the same spokes as that used in *Successfully Managing Change in Organizations: A Users Guide*. However, it is focused on the cultural aspects of the eight elements. The questions to complete the web diagram will be in the appendix and on the CD provided at the back of the book

Chapter 18 Assessment and Corrective Action

Having taken the cultural web survey and developed the diagram is only the first step. To successfully alter the culture the reader needs to conduct a cultural root cause failure analysis (C-RCFA). This will identify areas for improvement in multiple elements of the web and position the reader to make the necessary changes. This chapter will explain how this analysis is done.

Chapter 19 Moving Forward

Change is not a one-time event. If the organization is in a continuous learning mode, then it makes changes, evaluates the results, and positions itself to make further changes. This is the realm of continuous improvement or cultural evolution – always improving on what you have created. This chapter closes out the book with a brief recap of what was discussed and encouragement for the task at hand.

1.6 Navigating through the Book

Ideally this book should be read from front to back; the material is presented in a manner that provides you with information that acts cumulatively to build your understanding of organizational culture. This information includes the four elements that describe culture, the concept of vision and the Goal Achievement Model, the eight elements of change, and, finally tying it all together, a discussion of the Web of Cultural Change.

However, the reader may wish to jump around and read chapters out of order based on their immediate and specific areas of interest. For these readers I offer a road map of the book in Figure 1-1. All of the chapters are labeled as well as the way in which they tie together. This should provide you with the tool you need to navigate your way through the text.

1.7 Getting Started

As with any effort, getting started is often difficult. However, the value in a successful effort is well worth the time, energy, and commitment required. The reason behind this is that you are not just changing the way a process is executed, or a procedure is followed. By changing the organization's culture, you are in essence changing the very nature of the company. This book addresses this change of culture in a general sense that can be applied across many disciplines. But more specifically, it is targeted towards changing the culture as it pertains to reliability. If we are successful in this arena, the result will be a major shift in how work is performed. In addition, a culture that is focused on equipment reliability reaps other closely-associated benefits. These include improved safety and environmental compliance. Both of these are tied closely to reliability in that reliable equipment doesn't fail or expose a plant to potential safety and environmental issues.

Changing a culture to one focused on reliability also has spin-off benefits in many other areas. Reliability or the concept of things not breaking can easily be applied to other processes that are not related to equipment efficiency and effectiveness.

However, this type of change is very difficult. After all you are trying to alter basic beliefs and values of an organization. These are the behaviors that have been rewarded and praised in the past and may even be the reason that many in the organization were promoted to their current positions. Change may also seriously affect people's jobs because things they did in the past may no longer be relevant in the future.

I hope that you will find this guidebook useful, not just as a place to start or as a text to provide the initial information for your effort, but also as a book which will help you through all the phases of the process and into the future. I wish you success in your efforts and offer you this quote from Nicolo Machiavelli. Read and think about the message. It clearly describes the work you are about to undertake.

"There is nothing more difficult to take in hand, more perilous to conduct, or more uncertain in its success than to take the lead in the introduction of a new order of things, because the innovator has for enemies, all of those who have done well under the old conditions, and luke-warm defenders in those who will do well under the new."

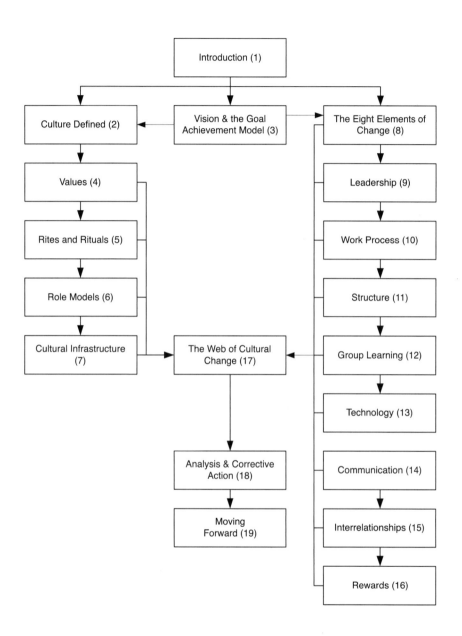

Figure 1-1 Book navigation

Culture Defined

Chapter
2

2.1 What Is Organizational Culture Anyway?

All companies want to improve and, where necessary, change the things that they feel will lead to better work processes, improved effectiveness and efficiency, and ultimately better profits. They go about this task in many and varied fashions. Often managers are replaced with those who senior management believe are more in line with what they want to achieve. Another strategy is to conduct a work process redesign, developing an "as is" process followed by the "to be" model. This redesign usually leads to restructuring, often with layoffs, and the implementation of new and different work processes. At times these companies often bring in high-powered consultants whose job it is to work with plant management to accomplish all of the steps I have described.

In my thirty-three years in industry, I have been involved in numerous initiatives designed to improve how maintenance was conducted. Often these initiatives brought with them statements that what we really needed to do was to "change the organization's culture." In most of these instances I was led to believe that those making this statement, senior managers or the consultants that they hired, knew what this meant. After all, these were the people making the "big bucks" so I had every reason to believe that they possessed this knowledge. In general I was wrong.

What I found out through inquiry was that these individuals often had no more clear knowledge of what it took to change an organizations culture than I did. However, in their defense, they knew that to be successful with new change initiatives there was this hidden force known as culture that had to be altered if the organization was to make progress.

Part of the problem with those who claim that they understand the concept of organizational culture change is that they have the same frame of reference as the rest of us. The senior managers have usually worked their way up from junior engineers through other jobs of increasing responsibility in the organization until they reached the senior manager level. The consultants who are hired to support the efforts have the same bias. If they didn't work their way up through the ranks of a business and then go into consulting, then they still progressed their career from the bottom to the top of their consultant firm.

In both of these cases what people learn along the way is what I will refer to as hard skills. These are skills like planning and scheduling of work, implementing a preventive maintenance program, and others made up of specific tasks that, when properly implemented, change the way work is conducted. These tasks and the change they bring are important; however, a majority of these initiatives end in failure. They fail when management "takes their eye off the ball" and moves on to other work. They fail when the sponsoring manager leaves and is replaced with someone who does not have the same passion for the initiative. They also fail at times due to open and active resistance from the organization.

What is missed in the training of most people, or if they receive training it does not receive the same value, is training in the soft skills. These are skills such as creating a vision, holding people accountable for their goals and initiatives, leadership, communications, and interrelationship building. To change an organization for the long haul and to avoid failure requires that these skills be employed constantly and consistently across the organizational landscape. Why? Because as shown in Figure 2-1, the soft skills are the foundation for hard skill implementation. We all know what happens if a structure is built over a poor foundation.

Therefore, in order to have a successful change of the culture we need to have an understanding of the soft skills and implement them before the "hard skills that we are trying to change. But this only touches the surface of that hidden force known as culture. You can implement the soft skills and even have them fail to properly function due to negative

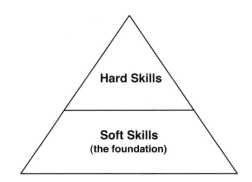

Figure 2-1 Hard Skill / Soft Skill Pyramid

cultural influence.

The seriousness of this issue is why people in the change business, senior managers or consultants, always are referring to the need to change the culture.

Years ago I was involved with the implementation of a Quality Program. It was one of those initiatives sold by a consultant that promised to improve the quality of everything we did in the plant. It had a strong soft skill component and this was followed by the introduction of the hard skills built on the soft skill foundation. The program lasted several years, but in the end it failed. We had attempted to change many things, but we failed to understand and change the culture. In the end, the culture worked invisibly behind the scenes to restore status quo to the organization.

The point is that many claim they understand organizational culture, but when you observe the initiatives being implemented and the failure of a majority of them, it should be clear that culture is not so easily understood nor do most organizations work to alter the culture to achieve success in their change programs. The real model we need to consider has culture as the sub-foundation of soft skills. If we can successfully alter the culture, then we can then build soft skill and later hard skill initiatives on top of it.

Now that we know where culture belongs in the change scheme, we need to get a clearer picture of what it is and how it interacts with the soft skills it needs to support. If we can achieve this picture, then we can successfully build the hard skills on top of the soft skills that are built on top of the cultural change. It is only then that we can achieve the level of change and improvement we seek.

Figure 2-2 Pyramid with culture as a foundation

2.2 Culture Defined

The starting point of our discussion is to define organizational culture in a way that is understandable. From this we can dissect the definition into its component parts and reach a clearer understanding of each. This should provide us with the starting point we need to discuss why it is important, the types of culture we need to learn to deal with, and how the work culture impacts the soft skills that we need to employ to make positive business change.

In his book **Organizational Culture and Leadership**, E. Schein defines organizational culture as follows:

A pattern of shared basic assumptions that the group learned as it solved its problems of external adaptation and internal integration that has worked well enough to be considered valid and, therefore, to be taught to new members as the correct way to perceive, think, and feel in relation to those problems.

If you think about this definition, it clearly describes that sub-foundation upon which an organization's soft and hard skills are built. It also paints a clear picture of how ingrained these basic assumptions are; this picture allows you to understand how difficult they can be to change. Let's look at the component parts:

"A pattern of shared basic assumptions"
The operative word here is that the culture is constructed upon shared basic assumptions. Because they are shared, when you try to change the assumptions you need to change them in everyone.

"The group learned as it solved its problems of external adaptation and internal integration"
The next part of the definition explains that these assumptions are not new creations. They have been tested over time as the organization learned how to solve both the internal and external problems that quite often were serious threats to their very existence.

"That has worked well enough to be considered valid"
Furthermore, these assumptions worked well for the organization, which has collectively considered them valid. Think about the problems you will face trying to implement change where the new initiative is in conflict with a basic assumption that has been vali-

dated over time.

"Taught to new members as the correct way to perceive, think, and feel in relation to those problems"

This last part locks the assumption into the culture because it is taught to all new members as the "way we work around here if you want to succeed."

This is a very powerful definition if you think about its far-reaching impact on new change initiatives. It essentially says that if a new initiative conflicts with a basic assumption that was learned over time, has worked well enough to be held as valid, and is taught to the new members so that everyone believes it as true, then changing things is going to be a very difficult task.

One of the most difficult changes to make in the area of reliability is to change a work culture from reactive to proactive. Let's examine some of the reasons that this change is so difficult in light of the definition.

Suppose our change initiative was aimed at implementing a planning and scheduling process that brought with it reliability-based repairs for every job. This implies that in our planning and scheduling process work is not scheduled until planned and ready. It further implies that equipment won't simply be repaired by throwing manpower and materials at it. Instead, the maintenance organization will take time to understand the reason for the failure, then execute reliability-based repairs so that the equipment doesn't fail again.

However, our new planning, scheduling, and reliability-based repair initiative is working against a culture that holds the following basics assumptions:

- Maintaining production is what is important
- Maintenance exists to serve production's needs
- When equipment breaks down it needs to be repaired as quickly as possible
- The maintenance crews need to be available to make repairs, not to do non-value-added work such as preventive maintenance.

How well do you think a reliability initiative would succeed in an organization with these basic assumptions? The answer should clearly be that it would not succeed at all. The new initiative is in direct conflict with what the organization believes to be true about maintenance work. Because they are still in business (at least for now), these assumptions have been validated and taught to new members. They exist as the sub-foundation of the business as described in Figure 2-2.

It should be clear to you at this point that change is very difficult to accomplish when it runs in conflict with an organization's culture. Yet it can be accomplished.

2.3 Why Is This Definition Important?

In *Successfully Managing Change in Organizations: A Users Guide,* I mentioned the three elements needed for successful change – a vision of the future, the next steps to get there, and dissatisfaction with the current state. The third element of these requirements for change indicates how you can overcome the resistance to change imposed by an organization's culture. Although I will explain this in detail in the balance of this book, an example will provide clarity.

Suppose we work in the company described above. They have a culture that values reactive repair, little or no planning, and a belief that maintenance simply exists to respond to production's needs-of-the-day. These are the basic assumptions validated over time. Suppose, however, that these assumptions no longer work, the equipment is always breaking down, and the company is losing money. In this instance, a change to a proactive work process is possible because one could challenge and clearly point out that the expressed value of reactive repair considered valid by the company is not valid. Such a challenge opens the door to new change initiatives and allows for the introduction of new assumptions. This dissatisfaction with the current state sets the stage for change.

2.4 Types of Organizational Culture

There are many types of organizational cultures; each one acts and inter-reacts differently. It is important to understand this if we want to be able to initiate a successful change program because, being different, each requires a different approach.

The differences in cultural type can be portrayed by a quad diagram – a diagram that compares two factors in a matrix format. In this case the x-axis is how the organization acts towards change. The two types are closed and open. A closed organization is slow to change and, in some cases, reluctant or even adverse to change. It believes that what has worked in the past will continue to work in the future. On the opposite side of the spectrum is the open organization. It is very open to change because it knows that to remain static is unacceptable if one wants to optimize or even stay in business.

On the y-axis is feedback. If you remember the definition of culture, it addressed the issue of organizations validating their group assumptions. It is through the validation of feedback process that the organization learns that its behavior is correct and appropriate. It learns this through feedback resulting from its collective behavior. The y-axis shows two types of feedback – slow and fast. Some process changes provide instant feedback. For example, a maintenance organization reacts to a plant problem, fixes the equipment, and receives instant praise from production for rapidly correcting the problem. Other changes, such as seeing the results of a preventive maintenance program, are associated with slow feedback. In these cases, you implement a preventive maintenance (PM) program and often do not see the results for a year or more when your failure-tracking metrics show a steady decline in the failure rate.

Taking these two factors – change acceptance (closed or open) and change feedback (slow or fast), I have developed a quad diagram depicting the various types of organizational culture – Figure 2-3

Let us examine the four types as identified by the numbering scheme in the figure.

1. Closed – Slow Feedback

This box represents change adverse organizations. They are highly conservative. As a result, everything that they do is overly analyzed. This process takes a long time and results in slow feedback regarding success of their change efforts. Coupled with slow feedback on any changes they introduce, they get caught up in analysis of the situation to the point that changes are never undertaken. Essentially these groups are in denial. They believe that what they have is what is best. Organizational progress is slow or non-existent, and failure is often the end result.

2. Open – Slow Feedback

These organizations are open to change and successfully deal with the fact that feedback from their change initiatives is slow. They set a vision, develop goals, initiatives and activities, and stay the course over the long haul. An example of this type of firm is one

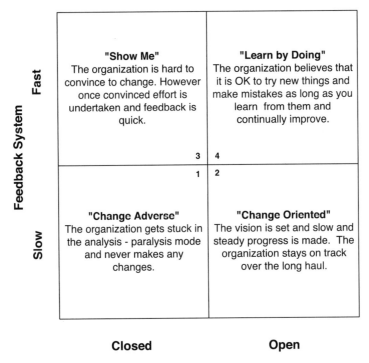

	"Show Me" The organization is hard to convince to change. However once convinced effort is undertaken and feedback is quick. 3	"Learn by Doing" The organization believes that it is OK to try new things and make mistakes as long as you learn from them and continually improve. 4
	1	2
	"Change Adverse" The organization gets stuck in the analysis - paralysis mode and never makes any changes.	"Change Oriented" The vision is set and slow and steady progress is made. The organization stays on track over the long haul.

Closed   **Open**

How the Organization Acts Towards Change

Figure 2-3 Cultural mode

that institutes a preventive or proactive maintenance program, then works hard over an extended time to make it work.

3. Closed – Fast Feedback

In this part of the quad diagram are the companies that are conservative in their approach to change, but when they do undertake initiatives, they seek rapid feedback. These companies need hard evidence that a change program will work before they are willing to attempt it. These are the "show me" firms. However, they do not wait to acquire 100% of the feedback from their efforts. They believe that once value is clearly demonstrated, they can safely proceed. Often this issue can be addressed by pilot programs to test a new idea and gain acceptance. The other aspect of this quad is that the feedback is quick. What this means for the change agents is that they will need to spend a lot of time convincing the company of the need for change. Once the effort has started, however, the rapid feedback will allow for rapid deployment.

4. Open – Fast Feedback

The last part of the quad diagram covers companies that are open to change and seek rapid feedback on the success from their efforts. These organizations recognize that change is sometime accompanied by failure. They are willing to accept that failure is a way of learning and are always open to attempting new things. An example would be a company that moves from a reactive work environment to one of planning and scheduling of the work. In this case, feedback as to the success of the planning effort would be rapid – the plans had value or they did not. In a "learn-by-doing culture" the successful parts of the process would be retained; those that were not successful would be discarded and new ideas tried out.

2.5 The Elements of Culture

In their book *Corporate Cultures: The Rites and Rituals of Corporate Life*, authors T. Deal and A. Kennedy describe the four components of a corporate culture – values, heroes, rites and rituals, and the cultural network. Their concept that organizational culture is composed of four key parts is a valid one; with some alterations, I will use the four-part model in my discussion of culture and how you can change culture in order to implement reliability-focused change in your organization.

Figure 2-4 below shows how each of these four components are a part of a larger whole which we refer to as our organizational culture. Each of these components plays a key role both independently and dependently as part of the cultural system.

Values

Values are the beliefs and assumptions that an organization believes to be true and uses as a set of guiding principles for managing its everyday business. They are what collectively drive decision making within a company. For instance, an organizational value may be that production is the only thing of importance and, when things break, they need rapid response in order to return them to service. Another example of an organizational value is that equipment should never fail where the failure was not anticipated through proactive maintenance work initiatives. Although these two examples are very different, in each case, the value

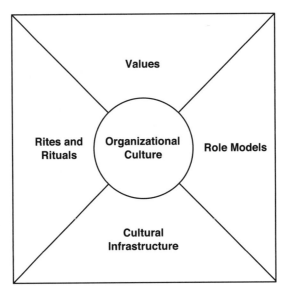

Figure 2-4 Four types of culture

described drives the collective decision making process for the organization.

Role Models

Role models are people within the company who perform in a fashion that the organization can and wants to emulate. They are successful individuals who stand out in the organization by performing in line with the corporate value system. They exist throughout the organizational hierarchy, from the reliability/maintenance manager through the highly-skilled mechanics. Role models show people that if you wish to be successful you need to follow the values set up for the organization. These role models are then copied by those who work within the business because they show how to perform within the culture. In addition, the role models are used as an example for newcomers to clearly show how to behave if you wish to succeed.

Rites and Rituals

Rites and rituals are the work processes that go on day-to-day within a company. They are so ingrained in how people conduct business that they are not actually visible to those within the company. Rituals are "how things are done around here." Rites are a higher level of rituals. These are the corporate events that reinforce the behavior demon-

strated in the rituals. For example, planning and scheduling of maintenance work is a ritual performed by the maintenance organization to
make repairs to equipment that has failed. The associated rite, the reinforcement of the ritual, is the fact that the organization requires a weekly scheduling meeting and adherence to the developed weekly plan.

Cultural Infrastructure

Cultural infrastructure is the fourth part of the organizational culture model. This is the informal set of processes that work behind the
scenes to pass information, spread gossip, and influence behavior of
those within the company.

These four parts, working in conjunction with one another, make up
that rather elusive thing that we refer to as organizational culture.
Each of these elements will be discussed in detail in subsequent chapters. The important thing to recognize is what people really mean when
they talk about cultural change. They mean that they wish to alter the
value system, displace people who are emulated, but are not in line with
the new values, change the rites and rituals, and reframe the cultural
infrastructure. Think about the implication of this change. It certainly
is a major step for any firm to take; which is why it is so difficult to
implement and make stick over the long term.

2.6 Sub-Cultures

We already know that in every organization there exists a culture
unique to that organization. What we need to discuss at this juncture is
that within each culture there are sub-cultures that are unique to individual departments or groups. These sub-cultures exist because of that
fact that every group within the company has specialized common problems that are faced only by those who are members. As a result, these
groups form sub-cultures that enable them to address these problems
and survive within their specialized environments.

These sub-cultures have unique traits but always include the domi-
nant or core culture of the business. The simple reason for this is that a
sub-culture without a foundation based on the business's core culture
could not long survive.

For example, take a machinery organization whose membership has
a value system that believes strongly in reliability-based repairs. This
culture supports the approach to equipment repair that requires an
analysis of breakdowns and the development of a repair strategy that

will return the equipment to service such that the cause for failure has been addressed and corrected. As a result, their leadership – the role models of this culture – provides the necessary time for the machinery engineers to do the proper level of analysis and development of corrective action plans – their rituals. This is wonderful in a reliability-based work culture; this machinery sub-culture will receive the support needed to be successful.

However, suppose this sub-culture is trying to survive in a reactive work culture. In this environment, production wants the equipment repaired and returned to service as soon as possible. There is no time for analysis and development of sound reliability-based repair plans. The directive is "fix it now and fix it fast!" How long do you think the reliability sub-culture would survive? In fact how long do you think the role models of the machinery group who were strong advocates of reliability-based repairs would stay employed? The answer is probably not very long. The reason is that the sub-culture is out of sync with the core culture of the business. This extinguishment of the sub-culture would also take place if the core was reliability-focused and the machinery organization was in a "fix it now and fix it fast" mode.

Sub-cultures, when they are able to exist, take on the same four traits as the core culture. They have a sub-set of values, role models, rites and rituals, and a distinct cultural infra-structure. It is logical to suspect that these sub-cultures will develop because departments or groups within a firm have their own set of issues; they need to develop

	Values	Role Models	Rites and Rituals	Cultural Infrastructure
Leadership	M	M	M	m
Work Process	M	M	M	m
Structure	M	m	M	m
Group Learning	M	M	m	m
Technology	M	M	M	m
Communication	M	M	M	M
Interrelationships	M	M	m	M
Rewards	M	M	m	m

M = Major interaction   **m** = minor interaction

Figure 2-5 Four Elements of Culture vs. The Eight Elements of Change

the sub-culture to be able to successfully address these issues.

When one group within a company is not as successful as the others, it is often because the sub-culture is out of sync. You see this quite often when a new leader takes over the organization or the organization is acquired by another firm. In each of these cases, a new culture is brought into the work environment. Some departments recognize the difference and adjust to get back into sync; others do not make the adjustment. Those that don't are the ones that usually get into culture / sub-cultural conflict.

2.7 The Eight Elements of Change

In my book *Successfully Managing Change in Organizations: A Users Guide*, I introduced the eight **elements of change**. These elements — leadership, work process, structure, group learning, technology, communication, interrelationships and rewards — are the key elements that if addressed collectively will enable a firm to undertake and be successful in implementing change.

However, as I learned more about cultural change in an organization, it became apparent that these elements each had one or more of the four components of culture embedded within them. This meant that not only were the change agents responsible for addressing all **eight elements** within their change initiative, but they needed to be very aware of how each of the eight elements impacted the **four elements of culture.**

Figure 2-5 below shows this relationship. The **eight elements of change** are listed vertically and the four elements of culture are listed horizontally. This creates a matrix that depicts all of the possible combinations in which the eight elements of change impact the four elements of culture. I have placed an "M" in each spot of the matrix where one the **eight elements of change** has a major impact on one of the four elements of culture and a small "m" where the impact is not a strong.

Let us look at one example so that I can demonstrate this relationship. Although this chapter only provides this one example, Chapters 9 through 16 discuss each of the elements in detail and show not only how they impact organizational culture, but also what you need to do with each so that you can effect positive change in your culture.

In Figure 2-5, the matrix indicates that leadership has impact on values, rites and rituals, and role models, three of the four elements of culture. Suppose that the plant in our example wants to implement a comprehensive reliability program. From a leadership point of reference, it needs to determine how to influence the culture and more specifically the three components in question if it wants the change to be successful. This can be accomplished by altering the present value system to one that reflects the change the plant wants to implement. Furthermore, the plant needs to put people into key positions that are reliability focused. These role models provide the plant personnel with visible day-to-day work performance targeting reliability as opposed to whatever model was in place prior to the change. Finally, the work processes or rituals need to be changed to reflect the reliability initiative. This can't be a one-time event, but rather a radical overhaul of the old process. The change of the rituals must be accompanied by a new set of reinforcing rites. In this fashion, we have addressed one of the eight elements and shown how, in order to refocus our work culture, leadership needs to be applied to three of the cultural elements.

If we don't apply the elements, what is likely to happen? In our example, the value system will remain unchanged, the processes or rituals that are how business is conducted on a day-to-day basis will remain unchanged, and quite possibly the people who remain in leadership positions will be those who are role models for the old processes, not the new.

This analysis can be carried out for each of the eight elements of change. The value that this delivers is twofold. First, it helps you think about the effort as it relates to the eight elements of change. This forces you to focus on the soft skills required by the change process. Second, it allows you to look at the eight elements of change in the context of the four elements of culture. By addressing change at this level, you do more than simply change the work process. You also change the culture that supports it, thereby delivering sustainable long-lasting change to your company.

Vision and the Goal Achievement Model

Chapter 3

3.1 The Need for Vision and Goals

It is said that "if you don't know where you are going, then you will never know when you have arrived." I would like to modify this statement and add an additional thought to focus it on the topic of change management. "As a leader within your company, if you don't know where you are going, then you will never be able to put in place processes that will get you there and you never will arrive." This statement makes two important points. First you need a vision to define where you want your firm to go. Second, you need a plan that will enable you to get from where you are to where you want to be.

Imagine a company that has in place a very reactive maintenance work culture. Its vision of maintenance is to arrive at work in the morning, find out what broke down over night, and then react by making repairs to the equipment and returning it to service as soon as possible. This firm's idea of a vision is to have a day when production isn't complaining about equipment in need of repair, how long it takes maintenance to make repairs, or the sub-standard quality of the repairs.

Now suppose you arrive on the scene as the new maintenance manager. You have recently been hired to bring to the company the techniques you used at your former firm which changed its extremely reactive work culture to one that was reliability focused. In addition, you helped to improve production by reducing breakdowns, and you provided effective and efficient maintenance services by addressing predictive and preventive maintenance strategies vs. the "break it – fix it" mode of reactive maintenance.

As the new maintenance leader, you have three tasks at hand. The first step is to show the organization that what it is doing, while getting breakdowns repaired and production back on line, is not an effective or efficient solution to the maintenance problem. This effort is not simple and is entirely focused on changing the work culture. I will hold off on this discussion for now because it is the topic that is addressed in the balance of this book. Instead, what I want to discuss are the other two tasks: creating the vision of the future (step 2) and providing the plant

29

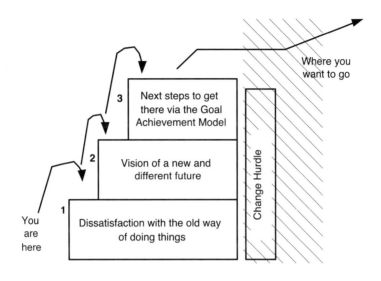

Figure 3-1 The change hurdle

with the next steps to achieve it (step3). Figure 3-1 shows the three steps you need to take if you want to get over the hurdle blocking you from successfully changing the work culture.

3.2 What Is Vision?

Many reading this book have set visions within their companies. I know I have on numerous occasions and with numerous managers. Before I define what I believe a vision is, let me tell you what I believe a vision is not:

It is not a long winded treatise

It is not a statement that is difficult to understand

It is not a tool reserved only for senior managers

It is not something fixed in stone never to be changed

Instead, a vision is a clear picture of something that the firm wants to achieve. It is short and to the point, understandable by everyone. The easiest way to tell if a vision is incorrect is to ask a cross section of the employees to state it. If everyone can paint the same picture, then the vision has been set and communicated correctly. Otherwise the process of setting the vision needs to be reworked.

In Successfully Managing Change in Organizations: A Users Guide, I defined vision as follows;

> Vision is an idealized picture of a future state, one that is integrated into the organization's culture. It provides a stretch, yet it is achievable over an extended time period with a great deal of work and collective focus by the entire organization. Because it continually evolves, it is never fully achieved.

Let's look more closely at this definition and discuss the component parts.

An Idealized Picture of a Future State =

Vision should be something that people can clearly see. When asked, all employees in the company should be able to describe the same end state – what their firm will look like when the vision is achieved. They may use different words, but the picture they paint needs to be the same.

Integrated into the Organization's Culture =

The vision must be difficult, if not impossible, to alter so that personnel changes can not easily destroy what the collective members have created. Too often the vision is not sufficiently integrated into the company culture. In these cases, a change in management can easily alter or destroy what everyone has worked to achieve.

A Stretch, Yet Achievable

The vision needs to be something that the firm can achieve, but not easily. If it is too difficult, people will become frustrated and give up. If it is not a stretch, then it will be easily accomplished and will not have significant value.

Extended Time and a Great Deal of Work

A vision is not something that can be accomplished in a short period of time. It represents a major shift in how a firm does business. If a vision can be achieved overnight, then it is not sufficiently a stretch for the firm. On the other hand, a worthwhile vision, one that takes a lot of time and requires a major shift in the culture, will take a great deal of work.

Collective Focus by the Entire Organization

It is not enough if just one person, or even a few people, under-stand and are working to achieve the vision. Instead, the vision must be a collective effort for the entire organization. Only then can it not only succeed, but also be long lasting and beneficial.

Continually Evolving, Never Accomplished

Although the vision is set at the beginning, the organization continually evolves; the end-state is never accomplished. In fact, by the time that the initially described end state is reached, a new and evolved end state will have replaced the original.

At this point, a few examples are in order. In 1961 President John F. Kennedy said "It is my plan to put an American on the moon and return him safely by the end of the decade." While this was many years ago,

Idealized picture of a future state	Safety programs are highly evolved in industry today. A picture of this level of performance related to reliability is certainly an idealized one.
Integrated into the culture	If a reliability program similar to the safety program is put into place, it will be integrated and not easily altered by new management.
Stretch yet achievable	The fact that a simialr vision can be achieved for safety clearly points out that a reliability program in the same format is a stretch, yet achievable.
Extended time frame	Programs of this nature can not be achieved overnight, so it fits the ex-tended time frame requirement.
Great deal of work required	If we reflect on the time and effort to improve safety, a similar effort for reli-ability will certainly take a great deal of work, require organizational focus, and total involvement.
Collective focus required	
Entire organization involvement	
Continuous evolution	Just as with safety, a good reliability program will continue to evolve as people learn from their efforts.

Figure 3-2 Comparing vision definition to a reliability example

Kennedy created a vision for the United States that was succinct and extremely clear. It put into motion a process that not only put a man on the moon by the end of the decade, but evolved into the space program that was responsible for the Shuttle program, missions to the planets, and other NASA-related achievements. All of this in only 21 words!

Let us take one more example that is related to plant maintenance and reliability. Suppose we set a vision for our firm that states, "We will operate and maintain our plant assets in a manner similar to how we address the safety of our employees." This vision again is very clear and concise. It is easy to remember and paints a very vivid picture of a new and different type of maintenance program. In Figure 3-2 we can examine how this vision fits the definition of vision stated above.

3.3 Why Is a Vision Required?

Without a vision, an organization has no destination or idealized future state. There is nothing to strive for and, quite often, the organization accepts the status quo as being the desired state that it seeks. In the world of maintenance and reliability, this view is acceptable if the status quo is based on a reliability-focused work culture. Although it is true that even in this culture there is always need for improvement, it certainly is better than an organization that believes the "break it fix it" mode is the work culture that it seeks. For companies without a vision of something better, there probably won't be any dissatisfaction with the current state and, as a result, no growth.

The next question we need to ask, if we agree that a vision is a requirement for organizational growth and cultural change, is who creates the vision of the future? The answer to this question is that the vision is created by the leadership. For reasons that we shall discuss in greater detail in subsequent chapters, it is up to the leaders to set direction – both short term and long term – and then work with the organization to make it happen. A very important part of this effort is to establish the vision for the organization and support the change required to make it work.

The second part – how do we accomplish the vision? – is not as easy a question to answer as it may seem. If the organization believes that the status quo is where it wants to be, then a vision other than what is currently in place will be difficult. If the organization is dissatisfied with its current state, it is headed in the right direction. However, because its collective frame of reference is the same and it shares a common set of experiences, it may not make the choice that will optimize its position in the future. So what is the answer?

In order to achieve a breakthrough change, an organization needs external focus. This can be accomplished in two ways; companies usually resort to a combination of both. The first way is to bring in new leadership. This step does not mean that the current leadership is bad, rather that the current leadership is restricted by their organization's culture. When outsiders are brought in to change a culture, they are often met by resistance. The strongest resistance usually comes from the incumbent leadership and role models of the old way or working. They are the ones with the greatest difficulty in seeing the need for the change.

The second way (which usually accompanies the first) is to utilize outside consultants. These individuals have a broad breath of experience. They have seen how companies that are experiencing similar problems have succeeded. In addition, they have seen how change can be accomplished in multiple industries and by employing many and varied methods. They also have experience facilitating change efforts – a very important ingredient for success. For more on how to effectively work with consultants, see my other text *Successfully Managing Change in Organizations: A Users Guide*. The answer that will enable an organization to create the vision it needs to be successful is to bring external influence into the mix.

3.4 The Goal Achievement Model

Once a vision has been created, the next step is to translate the vision into action. This is accomplished by the use of a tool referred to as the Goal Achievement Model. This model enables you to take a strategic concept and convert it into actionable work for the organization. By creating successive levels of detail, you move from a highly strategic concept to one that is tactically focused. The Goal Achievement Model has five parts – vision, goals, initiatives, activities, and measures. The first four are the main components whereas the last is the tracking tool for the model. Measurement will be discussed separately in Section 3.7.

Figure 3-3 shows the relationship among the first four elements of the model. In this figure, the x-axis represents the percentage that each element has as a tactical component. The y-axis portrays the same information for the strategic component. Using this model you can see the relative strategic and tactical percentages that each of the elements pos-

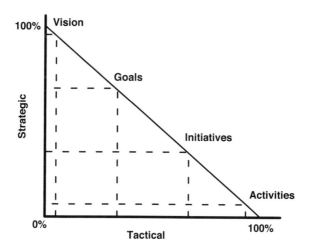

Figure 3-3 Tactical vs. strategic

sess. Note that as you become more tactical (moving from vision towards activities), you interact with subsequently lower levels of the organizational structure.

3.5 How the Goal Achievement Model Works

The Goal Achievement Model begins with setting the company vision. This is the cornerstone of the overall process. The next step is to identify several goals that support the vision. The relationship is one (vision) to many (goals). You need to be careful that you select only four or five goals. More than this and the organization becomes awash in too many goals and its efforts are diluted. Too few and the organization will run out of things to focus on as it works through the process. Remember that the goal stage is still more strategic than tactical so that goals are high-level efforts.

Next, initiatives are established based on the goals. Again this is a one-to-many relationship. If you developed four goals and for each goal you developed four initiatives, you would have sixteen initiatives. The initiative stage is more tactical and less strategic.

The last step is the development of specific activities. These are totally tactical in nature and are developed at the bottom of the organizational hierarchy. Again there is a one-to-many relationship. Extending our calculation, if we established four activities for each initiative, we would be working on sixty-four activities.

This example points out why we don't develop more than four of five goals at an one time. If there are more than five, it makes sense to complete one or more of them and then take on the others. If you try to address all of them at the same time, you spread the organization too thin and will probably accomplish very little.

As you can see in Figure 3-4, the significant benefit of using the model is that it creates a clear pathway for the organization to see how a vision can create goals which, in turn, can be used to create initiatives and ,finally, activities. There is a secondary benefit from the Goal Achievement Model: Those working on the details – at the activity level – can clearly see how the tasks on which they are working link upwards to the initiatives, then goals, and ultimately the company's vision. As a result, no matter the task, anyone can see how it supports the company's vision.

3.6 A Reliability-Based Example

Let us examine the Goal Achievement Model in more detail using a reliability-based example. The full model for this example is shown in Figure 3-5.

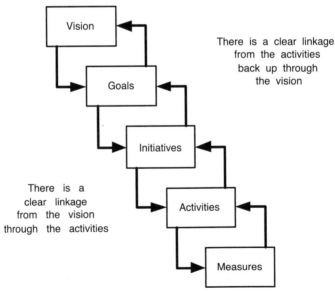

Figure 3-4 Goal Achievement Model – high level

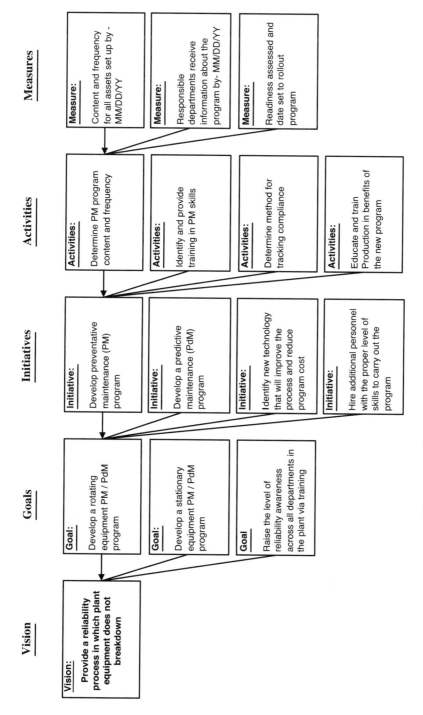

Figure 3-5 Example of full Goal Achievement Model

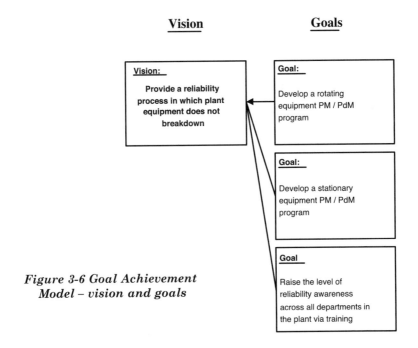

Figure 3-6 Goal Achievement
Model – vision and goals

Suppose that you plant had very poor reliability. Equipment was always breaking down and maintenance spent most of its time reacting to the failure-of-the-day. Resources were poorly utilized and profitability was suffering as a result. In order to improve this situation, management decided to implement a preventive and predictive maintenance program and set this as a vision for the organization. Management also believed that the Goal Achievement Model would be an excellent way to get everyone involved and part of the new process. With this in mind, the senior staff established goals based on their vision – Figure 3-6.

Continuing with the model, the middle tier of the organization was assigned the task of reviewing the goals. Then separate teams each took on one goal to further develop the initiatives and activities required to drive the work down through the organization.

The rotating equipment team took on the first goal: Developing a Rotating Equipment Preventive (PM) and Predictive (PdM) Program. At their first meeting, they developed several initiatives that they believed would help them develop and implement a program that would improve plant reliability. This goal from Figure 3-6 and the initiatives they created are shown in Figure 3-7.

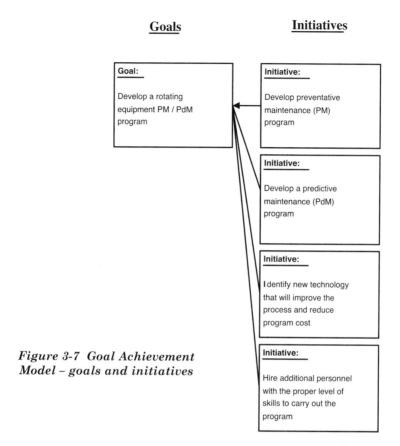

Figure 3-7 Goal Achievement Model – goals and initiatives

As we work our way down through the model, you can see that the work tasks get more focused and more specific to what we are trying to achieve.

Once specific initiatives are established, the next step is to engage those at the working level. The model cannot be used only for management. Everyone must be part of the process, working to achieve the vision. Regardless of their work levels, all contribute to the final outcome. With this in mind, the workforce is brought into the picture. Sub-teams are formed to develop and work at the activity level. For our example, we will select one of the initiatives: Develop the preventive maintenance (PM) program. Working on this initiative, the sub-teams developed the activities shown in Figure 3-8

At this point we have reached the bottom level of the Goal Achievement Model. It should be evident from Figure 3-5 that the benefit to management is the ability to clearly see a pathway from their

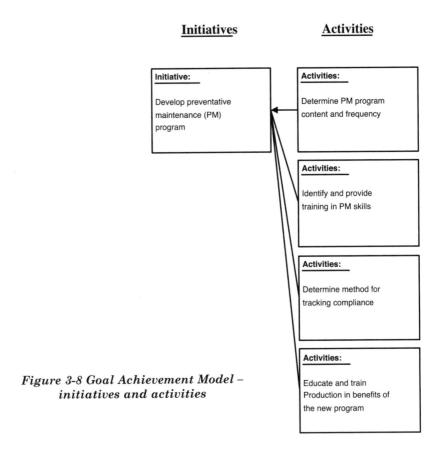

Figure 3-8 Goal Achievement Model –
initiatives and activities

vision all the way down to the activity level. Even more important is the fact that those working at the activity level (the workforce) can clearly see how their efforts contribute to the accomplishment of the specific initiative, which supports the goal, which in turn supports the vision.

This aspect of goal development and accomplishment has been left out of the equation all too often. It leaves the workforce unengaged in the improvement process and results in cultural stagnation. People will work very hard at all levels if they believe that they are adding value; conversely, if they do not believe that their work has meaning and benefits the company, they will generally not work hard at all.

3.7 Goal Achievement Model Measures

The discussion of measures requires additional clarity. Once management recognizes the model's value, they will want to put their own

Activities   **Measures**

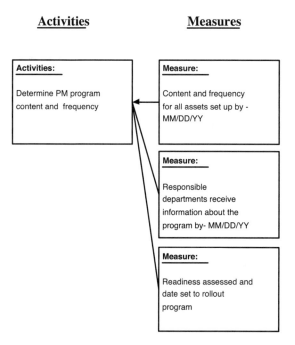

Figure 3-9 Goal Achievement Model – activities and measures

spin on how it is used and what some of the parts mean. I have seen this done most frequently with the last and most important section: measurement.

Managers work at the strategic level almost all of the time. As a result, their measurements are often at the same level. You may have heard these measures referred to as KPIs or Key Performance Indicators. At times these can be quite nebulous for the rest of the organization. People see the measures, but they are at such a high level that they do not have direct and immediate application for those at the working levels of the organization. For example, some possible KPIs are maintenance as a percentage of the replacement value of the plant, maintenance cost as a percentage of production, and others of the same nature. The problem with these measures is that most of us have no way of reconciling the measure with our day-to-day work activity.

Senior managers are going to want to create similar measures for the Goal Achievement Model. THIS IS NOT THE PURPOSE OF THIS PART OF THE MODEL. The measurement section of the model is designed to track activities. Using the first activity to determine PM program content and frequency, you can see in Figure 3-9 that the measures define when various aspects of this activity are going to be completed. This measure enables you to hold people who are performing the activity responsible and accountable for its completion. If I was developing the program content and frequency and I have to complete it by a specific date, then my manager should have a tracking tool similar to the one in Figure 3-9 to make certain this activity was completed.

3.8 Final Thoughts

As you will learn in subsequent chapters, cultural change requires you to address and alter key aspects that are at the very heart of the organization. Changing values, role models, and rites and rituals is a difficult task. However, the Goal Achievement Model can be used as a tool to support and facilitate the changes that you seek. It provides a way to structure the change and engage everyone in the process. Furthermore, if the measurement section is used as designed, it will enable you to hold teams and individuals both accountable and responsible for successful outcomes.

Organizational Values

Chapter 4

4.1 An Introduction to Organizational Values

Imagine that you are a foreman in a plant that manufactures a high-demand product. Your company works seven days per week, around-the-clock, to get the product out the door. Recently the plant's management has instituted a reliability program that has taken many of the mechanics who had been performing day-to-day maintenance away from this work. The program refocused them on preventive and predictive tasks. Also assume that this initiative was in response to poor plant performance reflected by high down time and lost production.

To make this scenario complete, imagine that you are a senior foreman, highly respected by your maintenance peers and production, to whom you provide service. The reason you are held in high regard is that over the years you and your crews have always been available to fix the "emergency of the day" and save the operation. Production has known it can always count on your team to react to its needs, pulling out all stops to get the job done.

But with the new reliability program now in place, you can't always be counted on as before. Several of your best "rapid responders," who happen to be the best mechanics, are off performing preventive and predictive maintenance tasks. Their absence has reduced both your crew size and your ability to respond to your customers' needs. The new program, which you never supported in the first place, has taken away valuable resources. Needless to say, you are frustrated, as are those in production who do not believe they will continue to get the response they need to keep up with the equipment breakdowns.

Today you are faced with the same dilemma you have repeatedly faced recently. A major piece of equipment has failed and production is screaming. Your choices are either to stop one of the less important routine maintenance jobs or to stop the preventive maintenance crew and divert them to the emergency. If you stop the routine job, one of your customers will be upset with your performance. These are the same people who praise you when you rush in and save the day. What do you do? - The choice is yours.

If you divert the preventive maintenance crew, as you have in the past, there will not be any repercussions because operation's main concern has always been with the daily production quota. They will be happy and, as a result, so too will your boss. Therefore, the answer is obvious. As has occurred numerous times in our own work environment, you divert the preventive maintenance crew, promising yourself that you will reschedule them to these tasks at the earliest possible convenience.

What you have just witnessed is one of the key aspects that define an organization's culture - organizational values. These values set the direction for the company and help people make decisions when faced with critical choices. In this example, the firm's values were to fix what breaks as quickly as possible to support plant production. The crews performing other routine tasks could have been diverted, but this would have violated the basic values held by those in the plant. Conversely, preventive maintenance can always be done at another time; no one has ever raised any objection when the preventive maintenance crews were diverted in the past. In fact, everyone is so wrapped up in the day to day, that you face a continual fight getting people assigned to preventive maintenance each week. The most visible evidence of the stated value of preferring a rapid response to emergencies over staying on track with the PM program occurs when senior management, those who created the preventive maintenance program, accept this behavior when it is brought to their attention.

Organizational values dictate how we behave on a day-to-day basis. Of even more importance, they dictate our behavior when we are faced with critical and, often time, sensitive business choices.

Suppose that the plant in this example placed high value on its preventive maintenance program. Suppose too that you were virtually forbidden to divert the preventive maintenance crews without senior management's approval, which was never given. What would your choice have been then for handling the emergency? The answer would have been to divert a crew from a less important routine task. Not only would this have resolved the problem, but production would understand that this choice had been made because the expressed organizational values emphasized the importance of preventive maintenance.

Examples of this type play out every day in our businesses because organizational values help us decide what is important to the business and how to behave. They are one of the key components of our study about organizational culture.

In this chapter we will explore the topic of organizational values in detail. This is an important concept because, as in our example, values dictate the thinking and decision making process of those in the company.

4.2 What Are Organizational Values?

What are your organization's values? Thinking about and identifying them may not be as easy as you think; the true values of a firm are not always written down. Instead, they reflect how the members of the firm collectively behave, how they conduct their business, and what they believe are the true measures of success. Nevertheless, take a few minutes and jot down what you believe are your firm's values. Next, have several members of your organization across multiple departments and at various organizational levels conduct the same exercise. If all of you reach the same set of answers, you most likely will have identified the organizational values of your company. It is my opinion that, except in rare circumstances, a single list will not be the outcome of this exercise. We will learn more about why this mismatch occurs throughout the balance of this chapter.

Our goal is to address the following questions:

- What are organizational values?
- What are their characteristics?
- How can we go about altering them if needed?
- How can we instill a new or modified set of values within the company?

Organizational values for our discussion can be defined as:

A company's basic, collectively understood, universally applied and wholly accepted set of beliefs about how to behave within the context of the business. They also describe what achieving success feels like. These values are internalized by everyone in the company and therefore are the standard for excepted behavior.

Think for a minute. Would you go out and buy a gun and then rob a store? (The answer should be no!) Ask yourself why you feel this way and you will discover that it violates a core value that you have been taught. It is an act against what you believe, what your family believes, and what is viewed as unacceptable behavior in our society. In other words it is a value that is collectively held by society.

Similarly, organizational values are similarly applied in a business setting. When faced with a problem, those within the organization will invariably make a decision that reflects the organizational values of that business. These decisions are often not made consciously because organizational values are internalized and taken for granted. When you make a decision supported by the values, you feel comfortable. When you don't, you sense that something is just not right with your world.

Take safety as an example. In many plants, safety is among the strongest organizational values. Numerous slogans, programs, promotions, rules, and regulations have been set up to govern our behavior. But all of these different efforts are in place only to reinforce the already existing organizational value that it is not acceptable behavior to allow someone to work unsafely.

Years ago I was a project manager on an operating unit that was shutdown for repair. We had a serious work delay associated with unloading catalyst from a multiple bed reactor. The project was behind schedule and the company was losing a great deal of money every day that the unit was off-line. At one point I was by the vessel opening talking with the contractor's employees. When I asked where the supervisor was, the men indicated he was inside the vessel. After about five minutes he emerged; I noticed he did not have on the proper protective equipment required by the permit issued for entry. He explained that he needed to see the workers in the vessel. He added that the required safety equipment was not needed based on the current state of the job, even though the entry permit stated otherwise. The organizational values of my company were not only that safety was important, but also that following the rules to maintain the safety of all personnel was of primary importance. As a result, I fired him on the spot and had him escorted from the plant.

The purpose of this example is to show one of many organizational values at work. Let's compare what happened to our definition and see how this event was affected by the value system.

Basic belief

Safety through training and continual reinforcement of the rules was a basic belief of everyone in the plant. A company employee would never have entered a vessel without proper safety equipment. Contractors who received comprehensive training should not have violated this rule either.

Collectively understood and universally applied

By continuously applying the same set of rules and regulations, everyone understood and applied them in their daily work. This should have applied equally to the contractor.

Wholly accepted way to behave

The rules were recognized by everyone as the only accepted way to behave. For those who did not follow the rules, the recognized negative outcomes included employee discipline, dismissal, or even worse – getting hurt or being responsible for getting someone else hurt.

What achieving success feels like for the business.

The company lost an additional day while the contractor flew in a replacement supervisor. Nevertheless, the company had such a strong safety-oriented value system that it made clear to everyone it would rather lose money than have someone get hurt.

Internalized

In my case, as with my immediate supervisor, the safety value system was well internalized. When confronted with this situation, I did not hesitate; I shut down the job and fired the supervisor. Upon hearing of this action, my supervisor started screaming, not at me, but at the contractor for violating the safety regulations. Having worked with us for along time, the contractor should have internalized the rules as well.

Organizational values also apply in the area of plant reliability. If the values of our foreman and the plant management in the previous example were reliability-focused, then the question of diverting the preventive maintenance crew would never have crossed his mind. Conversely, when confronted with a rapid response issue, the collective values said that stopping a job for a customer – even one of low importance – was unacceptable, but diverting the preventive maintenance crew was well within accepted behavior.

4.3 Written vs. Actual Values

I am sure that each one of us has experienced the written company value statement.These are often the long written-out documents that are posted on bulletin boards or sent home to us in the mail from the corporate office. Many of these statements represent in fact how the firm at the shop floor level performs, but often this is not always the case. Often the written word describing how those who run the business perceive we should behave runs contrary to what actually takes place. Some examples of this contradiction are show below.

Written	Actual (Expressed)
Safety is our number one concern	Safety violations are accepted to get the job done
Reliability of our assets is of primary concern	Poor or non-existent reliability program and no enforcement of compliance where one does exist
Strict rules for working with contracts and contractors	Those who prepare contracts do not always follow guidelines
It is okay to stop a job if the employee feels there is a safety- or work-related problem	You better not do this unless you want a permanent assign ment to the night shift
Equal consideration for promotion and acceptance of the most qualified person	Hiring of friends and relatives often without the required qualifications

Figure 4.3 Written vs. actual values

In these cases, the daily dilemma that we face as employees is selecting which set of values to follow. The answer is that ultimately we follow those values that are in line with the actual behavior of our work society. We follow the unwritten rules, not the ones posted on the corporate bulletin board. The main reason for this is that the written values are ideal, but not always relevant to the day-to-day performance of our work. Evidence is around us all the time because we see how managers, peers, and the workforce act and react in real life. Furthermore, if we were to follow the written word (when it is different from what actually takes place) we would be brought back into line by peer pressure. We would be instructed on how work really gets done around here.

A wide discrepancy often exists between the written and the actual behavioral values within a company. When both are in alignment, a company or plant can be considered healthy from the standpoint of organizational values and work culture. In a later section I will address the issues around a value system in need of change.

4.4 Values Gone Astray

We shall examine four areas in this section so that you will be able to recognize value system problems at your plant. In each case, the values of the organization have gone astray. The areas are: 1) values that are no longer relevant or are obsolete, 2) values that are inconsistently applied, 3) values that fail to match organizational reality, and 4) values that meet resistance from the organization.

No Longer Relevant or Obsolete

Values that are no longer relevant need to be changed. However, the real question that needs to be addressed is how do we know when values that have served us well for years are no longer relevant to our business?

Take, for example, reactive maintenance. For years this was the standard mode of operation for a great many plants. It was an accepted value. It developed from a combination of poor equipment reliability, inexpensive manpower costs, and the high cost and business impact of equipment being unavailable for production. Plant maintenance forces became expert at rapid response or, as it was commonly called, "fire fighting."

Over time, manufacturing equipment became more reliable and technological tools to predict equipment failure improved. The result was that the strategy of fixing things that broke was replaced by a strategy that targeted predicting failure before it happened. This approach was far more cost effective. Yet the rapid responder value has continued to thrive in many plants. Clearly this value has outlived its usefulness. The way we can recognize this is that the new value, in this case, reliability-based repair, enables us to perform in a much more effective and efficient fashion.

Inconsistently Applied

Another area where values can create problems occurs when they are inconsistently applied. Inconsistent application sends mixed signals to the organization. Standardized behavior with regard to organizational values is essential because it provides guidance, direction, and support for consistent decision making.

Let us look at the example of emergency maintenance. Many plants define an emergency job as one that threatens production, the safety of the employees, or that of the surrounding environment. These jobs are typically worked until completion or until the true emergency aspect of the work is resolved. This approach is a value of the company, indicating clearly that it will take the proper action to correct emergency problems. However with limited resources, production teams frequently label jobs emergencies so that they can divert resources to their problem-of-the-day, away from other areas where they are working. Then when the day tour comes to an end and the maintenance organization is looking for authorization to work overtime, the work is no longer an emergency. If management allows this practice to happen, the value of emergency response will be eroded because of inconsistent application.

Fail to Match Organizational Reality

In many cases, a mismatch develops between the expressed value of the company and what actually happens on the shop floor. This area is similar to inconsistent application but worse because in this case the company's stated values are not followed and everyone knows it. Inconsistent application of the values can be corrected through closer monitoring, but mismatched values destroy the credibility of the firm. Because everyone knows the value is hollow, they ignore it. This problem can then spread to other values and required behaviors within the business.

For example, a company states for the record that safety is its most important value, taking precedence above all else. Yet the company's safety record is deplorable, it has no real safety program, and everyone knows that safety takes second place to production. This obvious mismatch of the value system with actual behavior tells people that their leadership doesn't "walk the talk." The result is confusion and a loss of credibility. Remember that values are there to guide the decision-making process of the group. What happens to this company when employees are confronted with a safety issue that will cause a loss of production? Some will interrupt production to be safe – the expressed value, yet others may not and the result could be catastrophic.

Meet Resistance

The last area where values have gone astray is when they are resisted by the organization even though they are the correct thing to do. In this case, something is happening within the organization to cause the resistance. This area is the most difficult one to address because resistance is not the root of the problem. Instead, it is a symptom of something else that needs to be addressed. If there is resistance to replacing an obsolete value with something better, why would people not want to change for the better? If the resistance is tied to efforts made to consistently apply values or to correct mismatches, why would people fight alignment? After all, alignment would help them make the correct decisions when faced with numerous choices.

The simple answer is that people do not like change. It removes them from their comfort zone and puts them in unfamiliar and uncomfortable territory. If resistance is to be overcome and the new values are to be institutionalized, then time will be required for people to readjust their comfort zones.

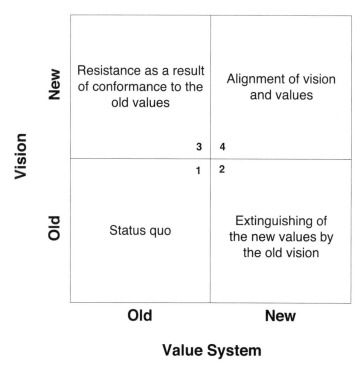

Figure 4-1 Vision vs. values

4.5 Vision and Values

Vision and value systems go hand-in-hand. As shown in Figure 4-1, without proper implementation and changes to the value system, a new vision will run into trouble. Similarly, trying to institute a new value system within an old or outdated vision will have problems.

Quadrant 1 – *Old Values and Old Vision*
This quadrant represents the status quo. The old vision is supported by an equally old value system. The combination may work for now, but problems are on the horizon. As the competition improves, a company in this quadrant doesn't.

Quadrant 2 – *New Values and Old Vision*
This quadrant represents the case of trying to change the value system within an old or outdated vision of the business. In this case, the internal cultural pressures will defeat the new value system regardless of the benefits that may have been included. This scenario will be discussed in Section 4.7.

Quadrant 3 – *Old Values and New Vision*
In this quadrant, a new vision is attempted without taking the necessary time and effort to change the values. This case of the vision and the value system being out of alignment will be addressed in Section 4.6.

Quadrant 4 – *New Values and New Vision*
In this case, the vision and the values are in alignment. They support each other and with proper implementation successful change will be the outcome.

4.6 When Values Are Out of Alignment with the Vision (Quadrant 3)

The old organizational values are out of alignment and need adjustment when they are in conflict with the company's vision. This isn't always obvious because values are not something we constantly think about as we go about our jobs. Values are deep within us. They govern how we work and, as stated earlier, they are there when needed to make the critical decisions that make our businesses successful.

How do we know that our values are out of alignment? The answer is when we see that the goals, initiatives, and activities associated with the Goal Achievement Model do not seem to work or reach successful conclusion. The work of the Goal Achievement Model is to promote the vision throughout the organization so that each level is engaged and can clearly see its contribution. When this isn't working, there is clear misalignment between the vision and what people are doing. Because the values of the organization essentially dictate how we make decisions, the conclusion is that the vision is out of alignment with the values.

Years ago I helped facilitate a work redesign from a reactive maintenance process to a work process in which the work was planned and scheduled on a weekly basis. In the reactive mode, the organization assigned planners, foremen, and work crews to each of the several production areas. The planner job was essentially to assist the foreman as they went about the day-to-day work. Planning was non-existent. The foreman and the work crew were there to respond to the daily needs of production. Because there was only minimal planning, work was never ready for the crews. It was never materialized properly and a considerable amount of productive time was lost.

The new process assigned a planner to each work area, but their new job was to actually plan the work. Once the plan was completed, production selected the work for the next week. That work was then scheduled and work crews were assigned from a central pool. The foremen and the crews no longer worked in a specific area, they worked on the important jobs, regardless of where they were located in the plant.

This was a new and different vision for both production and maintenance. We thought we had created a process that would severely reduce reactive maintenance because we had taken away the ability of the foremen to respond in this manner. We were wrong!

We changed the work process and created a new maintenance vision, but we did not change the core values which were:

- Production felt severe discomfort if any of its equipment was having a problem and not being repaired quickly. The reason behind this discomfort was a history of punishment for production loss and a lack of confidence in maintenance.

- The maintenance foremen felt severe discomfort because the tie with their production area had been broken. They no longer felt committed to any one area because they were never in any area long enough to feel committed.

- The former reward structure was missing for both organizations. Production managers did not receive praise for getting the pump that was not functioning in the morning running by the end of the shift. The foremen did not receive praise from the production counterparts for saving the day.

As a result of the work process change, we had taken a great many people out of their comfort zone. We had caused a major misalignment between the new work process and the existing values of those who we were asking to follow it. The organization reacted in a way that restored alignment with their value system.

- The level of emergency work went up as production forced response to the crisis-of-the-day (whether or not it really was a crisis).

- The few unit mechanics that were assigned to the operating areas for minor jobs began working jobs that were far from minor.

- The foremen working a planned job would divert their crews to other work requested, without approval to change the plan.

- Management failed to address these issues, indicating by lack of action that what was taking place was acceptable.

Needless to say the work process change failed as everyone continued to work within the value system under which they had worked most of their lives.

The scenario above takes place quite often in many of our businesses for the exact reason cited in the example. What can we do about it? As a first step, a new value system must replace the old. In that way, when confronted with critical decisions, everyone will not choose the old way but rather the new. Easily said – not easily done.

To correct the problem in the example, we would have to understand the existing value system and determine how it was not aligned with the vision. Once these areas of misalignment were identified, we would need to develop process changes to address them.

4.7 New Values and an Old Vision (Quadrant 2)

A new set of desired organizational values may be exactly the right thing for your business. However, there still will be a problem if the vision is not changed first. The reason for this is that the vision, often expressed as core beliefs, is reinforced in many and varied fashions. Over time, those values that are in conflict will be eliminated. Therefore, the vision must be altered first; then the values can be brought into alignment.

Suppose that senior management strongly believes that maintaining production levels is the only thing that really matters. The message that the plant receives is to maintain production at any cost. This message has led to unsafe work practices and people taking undue risk with the plant assets in order to keep up the level of production. This company has had numerous cases of safety, environmental, and reliability failures because the workforce made the wrong decisions about how to operate based on a set of wrong values.

Nevertheless, the vision of maintaining production at any cost permeates both the work system and the value system of those involved. Groups and individuals who hold other values, such as proactive maintenance where equipment not breaking is preferable to the "break it – fix it" mode of operation, are quickly corrected and refocused. If they persist, they are replaced with those who hold the same values as the rest of the organization.

As long as the vision and core beliefs of the senior managers remain as "production at any cost," the value system will remain in place to support it regardless of the fact that it may be incorrect. The conclusion from this brief scenario is that, unless some major event occurs, the existing vision and supporting value system will remain and the new values in conflict will not survive.

Continuing with the example, as a result of the production-at-any-cost philosophy, people were hurt and the company lost money. Ultimately it was sold (the major event). The new owners had a different vision. They believed that safety of the workforce, protection of the environment, and operating the equipment reliably would create a firm that would be both profitable and successful in the industry.

This new vision has moved the company to Quadrant 3 - the value system currently in place is now out of alignment with the new vision. This is the scenario described in Section 4.6. Figure 4-1 has pointed out that to improve, the vision must change first and the values follow.

Making change in the reverse order allows the old vision to extinguish the new values.

4.8 How to Change

Changing values to bring them into alignment with the vision is not an easy task. People have lived with the old set of values, which have been internalized. These values have served people well over time; often their performance within the old value system may be the reason they were promoted, retained during a layoff, or even highly respected within their workplace.

The new value system must reinforce the vision that you are trying to achieve. It must be created in such a manner that the organization makes decisions based on the new values, not the old. Additionally, you need to understand that something that has grown and been reinforced over years is not changed in an instant. Nevertheless, changing values is not an impossible task. The steps to follow include:

Step 1 – *Develop a clear and concise vision of the future.*

Because values will develop to support the vision, developing a clear and concise vision is critical. In addition, the vision needs to be one that will stand the test of time with only minor adjustments. This task must be driven be senior managers. The vision must be an expression of their core beliefs and future direction for the company. Suppose that part of an organization's vision stated, "We will operate our plant in a manner that treats reliability of our assets the same way we treat the safety of our employees." One can only imagine the values that would be developed.

Step 2 – *Communicate the vision continuously to the organization.*

Once the vision has been created, it needs to be communicated. Most people think that this means holding a meeting or a training program and explaining to everyone the new vision. That is a wrong assumption! Communication in this case is far different. Although presentations and training are required, what is ultimately needed is for the behavior of everyone, starting at the top, to change. In this case, communication is achieved through the visible behaviors of those who are designated as leaders. Management must "walk the talk." If they do not, the organization will cease to believe that the vision holds any value and will instead revert to the old way of doing things. How long do you think values associated with improved reliability will last if production continually forces maintenance to return equipment to service without allowing reliability-based repairs to be instituted?

Step 3 – *Build goals, initiatives and activities in support of the vision*
A vision can never be turned into reality without goals, initiatives, and activities to support it. This is the Goal Achievement Model, which was discussed in Chapter 3. (It is discussed in even more detail in *Successfully Managing Change in Organizations: A Users Guide* Chapter 5.) The Goal Achievement Model enables the vision to have an effect on everything that is being done in the plant and helps everyone to recognize this fact.

Step 4 – *Hold people accountable to the work they are doing in Step3.*
Holding people accountable for accomplishing the goals, initiatives, and activities associated with the Goal Achievement Model in Step 3 is critical. As soon as the leadership takes its attention away from these value-changing tasks, then progress and the tasks themselves have the potential of slowing down or even coming to a complete halt. The process needs constant reinforcement and attention. This is not easy to do because a great many other things pull the leadership's attention away from this effort. However, without constant attention and accountability, the effort to change the value system will not be successful.

Step 5 – *Reward those in alignment and punish those who are not. Maintain zero tolerance for deviation.*
This step goes hand in hand with Step 4. Zero tolerance for deviation means exactly what it says – deviation is unacceptable and brings with it severe penalties. This step needs to be expressed in action, not just words. When the organization sees that its leadership is serious, the values will begin to change. If safety of the workforce is an expressed value and a serious violation results in termination, the organization quickly will realize the consequences of violating the new value system.

4.9 Success Looks Like

Changing an organization's vision and then working via the Goal Achievement Model to change its value system are certainly very difficult tasks. Yet they must be addressed if we truly seek long-lasting change and improvement for our reliability and maintenance work initiatives. Although success may be difficult to measure, you will recognize it when those in the organization instinctively make decisions that support the new reliability-focused values vs. the former "break it – fix it" mode of operation

Role Models

5.1 What Is a Role Model?

When I was 29 years old, I was promoted to the position of Zone Supervisor. A zone supervisor was responsible for all of the maintenance work within a portion of the plant. In this job I had responsibility for approximately eight foremen and one hundred mechanics. I was the youngest person the company had ever promoted to a position of this nature. There were many reasons why. First, I had been successful in all of my previous positions and, second, I was part of an aging workforce. Two others and I were the first management employees hired into the maintenance organization in 20 years; everyone I worked with was over 50. The organization was trying to develop younger managers who could take over when the current managers retired.

I was very aware in this new role that I wasn't clear what I was to do. Although my organization was older and far more experienced than I was, I was expected to lead them and manage their work. My predecessor had been very successful and was highly respected by the entire organization. He had the ability to motivate people and get things done regardless of the circumstances. He had not been moved to another position. He had been given several special projects, one of which was to teach me the ropes. At the time, our organization was based on a reactive work culture. When things broke down, it was maintenance's job to fix them as quickly as possible and return the operation back to normal.

Being young and inexperienced, I copied the former supervisor. On many occasions, I asked for his opinion, help, and support in different circumstances. Over the next year, we had a very good relationship and he taught me what he thought I needed to be successful in my new position. In short, he was my role model. He had shown me how to be one of the best reactive maintenance professionals in our organization and shortly thereafter I was again promoted.

Several years later, the plant was sold to a private owner. At the time, I was in charge of all maintenance work in the plant. I reported directly to a maintenance manager brought in by the new owner to implement a reliability-based work culture. My specialty was reactive repair, not reliability. However, I quickly took on a new role model and learned that there was more to work than fixing broken equipment. My new role model taught me about the concepts of reliability, good planning, and scheduling techniques as well as how to implement programs that (with production's help) avoided equipment failure. The manager was successful in his conversion of the business and, as a key part of his team, so was I.

In both of these instances I emulated someone who I believed would provide the best maintenance services to the plant in the existing culture. Although my role models exhibited different traits and behaviors, they were correct for the culture in which they operated and were successful in their careers.

The remainder of this chapter will examine role models, why they are or are not successful, and how and why people emulate their behavior. We will also examine how and why people's behaviors, and frequently their beliefs about how to operate the business, can be altered by these models.

5.2 Role Models Defined

Role models are people within organization who exhibit traits that appeal to us and which we can apply to how we conduct our business. These role models are usually at or near the top of the organization; they have been successful within the organizational culture. They demonstrate a successful behavior style within the business culture, one that we feel comfortable adopting as our own.

Let us discuss further the three key components of a role model.

Top of the organization

Most people who are used as role models are at or near the top of the organization's hierarchy. These are the people we view as the most successful. They are the managers of our departments, the leaders, the ones who set the direction for the business. The key word here is success. Because those at the top are perceived as successful, we tend to use them as role models.

There is another reason why we often choose our leaders as role models. They set the expectations of what we are to accomplish at work. In most cases, these expectations are in line with their expectations for themselves. As a result, we emulate and assume their style because we are all working towards the same end. In addition, failure to achieve these expectations usually has severe negative outcomes. Therefore, modeling the manager to achieve the desired results makes sense.

Successful within the organization's culture

The second component is that role models are not just successful, but they are successful within the existing culture. This is very important. Think about how I used my former managers as role models. In each case, they were successful in their respective cultures and, therefore, were good models for me. But what if the situations were reversed? Suppose the manager who was reliability-focused was placed in a reactive maintenance work culture. How well do you think he would have succeeded? Who would have wanted him as a role model?

A style we can identify with and adopt

Even those some people are successful within the culture, there still may be reasons why we would not choose them as role models. If we truly want to use people as role models, we need not only to view them as successful, but also to feel comfortable adopting their style of management.

Suppose you are the type of person who firmly believes that all people within the workforce have unique value and should be treated with dignity and respect. Further suppose that your manager (who is a successful part of the organization) has achieved this position by acting and behaving in exactly the opposite fashion. Could you accept this person as a role model? Your answer would probably be no. Although you want to behave in a manner that will provide you a successful career, the behavior of your manager could never fit your personal beliefs and manner of conducting business.

5.3 Strategic or Tactical Alignment

From my previous examples, you can see that role models are aligned with the culture in which they exist. Reactive role models succeed in reactive cultures as do proactive role models in proactive cultures.

Role models are also strategically or tactically focused regarding how they conduct their work and how they support change within the work

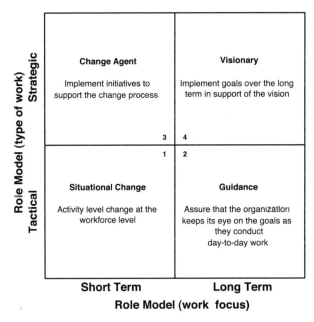

Figure 5-1 Role model – type of work vs. work focus

place. Furthermore, their work is often either short term or long term. This comparison can be best viewed in the quad diagram shown in Figure 5-1.

In this figure the type of role models (strategically or tactically focused) is depicted on the y-axis. Their focus – short or long term – is shown on the x-axis. In this way, we can represent the different components and discuss the role each plays in changing an organization's work culture.

Let us review this quad diagram from the perspective that we want to change from a reactive to a proactive, reliability-focused culture.

Short Term – *Tactical (Quadrant 1)*

Role models who fit Quadrant 1 work on the front line; they are viewed by their peers as the best at what they do. They typically have a day-to-day tactical focus and know how to get things done no matter what problems confront them. In most plants, these role models are the best at firefighting and reactive repair. However, we want to change to a more reliability-based model, and they need to change their focus. This can be accomplished, but not without difficulty. As you learned in Chapter 4, the organization's values must be reliability centered. You will also see in Chapters 6 and 7 how rites and rituals as well as the cultural infrastructure can make this change a difficult task.

Long Term – *Tactical (Quadrant 2)*

The role models found in Quadrant 2 do day-to-day work, but are more focused on longer-term goals. Typically these are the field superintendents – the people who manage the foremen and field work crews and who have the responsibility for leading their portion of the organization.

Short Term – *Strategic (Quadrant 3)*

The role models in Quadrant 3 support the line organization in a staff capacity. They include planners and schedulers, engineers, and possibly even consultants hired to implement reliability improvement processes. They are not directly involved in the day-to-day effort, but their work direct influences it. They keep the strategic initiatives of the organization in front of those immersed in the day-to-day work activity.

Long Term – *Strategic (Quadrant 4)*

Quadrant 4 is filled by the organization's managers. Their responsibility focuses on the longer term strategic goals and with those in the other quadrants to accomplish them.

Figure 5-2 shows how the quadrants described fit with the Goal Achievement Model, as discussed in Chapter 3. The quadrants from

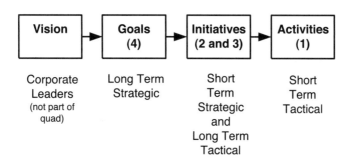

Figure 5-2 Strategic vs. tactical role models

Figure 5-1 are also shown in brackets.

Different groups and people select their role models based on where they work within the organization. Those on the line may select short-term models, either tactical or strategic, because the majority if not all of their work is short term in nature. If you were a foreman or mechanical engineer trying to implement a preventive maintenance program in the field, you would probably look for a role model in quad 1 or 3. These people are the ones who have been successful getting similar things accomplished within the culture.

Conversely, if you were trying to develop a strategic direction for the business such as self-directed work teams at the maintenance level, you would select role models who were successful in quads 2 and 4. These individuals have been successful in developing and implementing a vision and supporting goals for cultural change.

5.4 Cultural Alignment

Because this book is focused on changing from a reactive maintenance work culture to one that is reliability focused, we need to discuss what happens when the new role models are introduced. This is a real problem when trying to implement a reliability-focused work culture because invariably those who have been successful in the existing culture are usually not of this mold. Remember my example where my initial role model was highly reactive in a reactive culture and then a new plant management team was appointed with a reliability focus. In my case I made the transition, but what if I hadn't?

A role model cultural mismatch can occur at every level within the organization. For example, a company is purchased and the new management team has different plans for the business than did the former. A new manager is hired with a different outlook. A new superintendent is promoted who decides a change is needed, and new foremen are hired who are not content with the status quo. In each of these cases, those in the lower tiers of the organization are confronted with new role models and a new set of expectations that may be far different than those of their prior managers.

Everyone at some time (and often more than once) will have this experience. Owners, managers, and supervisors change; the role you need to model for success often changes as well. You have two choices. First you can adopt the new cultural role model as long as it doesn't conflict with your basic beliefs. This will maintain your position and you may learn something along the way. In my example, when my new manager introduced reliability concepts to my organization, he opened my eyes to a new and better way to do work. Although the change meant a new set of concepts to be learned and applied, the value to the company and also to me was immense.

The second option is to leave. If the new expectations are that much in conflict with your work processes, then leaving is a good option for you and for your firm. Otherwise neither will be happy over the long term. Admittedly, leaving is easier said than done. However, if you stay and do not adopt the new model, you will be viewed as someone who resists the change. Ultimately this image will make the work place difficult for you and all of those around you.

There are those who also believe that, if they wait long enough, the new culture and those associated with it will move on and the program will fade away. Although this may be the case with individual projects, it is not the case with a major cultural shift such as one from reactive-to reliability-focused maintenance work. These changes are major; once you begin moving down the reliability path, there is no going back. As someone once told me, "The train is leaving the station. Get on board for the ride or get off."

5.5 Bad Role Models Have Value

One other type of role model is worth discussing - one which is the most difficult to work with in our jobs. This is the person whose beliefs and actions are so opposed to our own that it is virtually impossible to adopt his or her style of management or behavior without violating who we are. There are alternatives when you are confronted with this type of situation. You can leave the organization and seek work elsewhere. You can attempt to transfer to another department. Or you can try to stick it out and survive, hoping that the individual will leave before you do.

At one point in my career, I worked for such a person. He was not what I considered a good manager. Nor did he treat the people who worked in the department with very much dignity or respect. To make matters worse, to run the daily maintenance operations he hired a person who had the same style of management as he had. This was a very difficult time in my career. I could not transfer and chose not to leave, but I also decided that I would continue to function as I had in the past using those who trained me as my role models. This dilemma is similar to the one described in the previous section except I was not being asked to change and model a better behavior. In choosing to reject the role model of my manager, I often got into difficulty. Fortunately for me, the individual he hired was promoted to another plant and the manager retired shortly thereafter.

What I learned is that not everyone is a positive role model. We are often presented with what I will refer to as "good bad examples." These are people who we can look at and say "here is someone who I do not wish to act like." If you examine why you feel this way and adopt behaviors that are opposite and more in line with how you feel you should behave, then they will have done you a great service. They will have shown you a model that you will choose to reject for a more positive (and opposite) behavior.

5.6 Role Models Are Created NOT Born

Thus far we have assumed that our role models are proactively focused, the ones who will support the change in culture from one that is reactive to one that is proactive and reliability-focused. However, what if discover that the predominant role models throughout the organization are not those who support change, but instead those who support the status quo? An even more challenging scenario: What if proactive role models don't exist at all in the organization? This is not an unrealistic expectation. Those who advocate proactive maintenance don't survive long in a reactive work culture where success is viewed as being the best "fire fighter" you can be.

Figure 5- 3, which illustrates a reinforcing loop explains the problem faced by most reactive work cultures when it comes to developing role models who support a different type of work culture. In block 1, things break. Equipment breaks down and production is interrupted. Maintenance responds in block 2, making the quick fix and returning the equipment to service. In this environment, the fix is not reliability-based; there is no time to discover what really went wrong so that action could be taken to prevent reoccurrence in the future. As a result of this type of work culture, those who made the quick fix are praised for "saving the day" and rewarded accordingly – block 3. This behavior is often a driving force for promotion. Block 4 has the balance of the people in

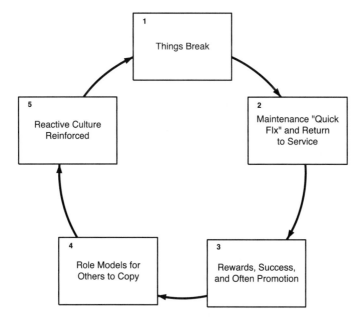

Figure 5-3   Example – reactive culture reinforced

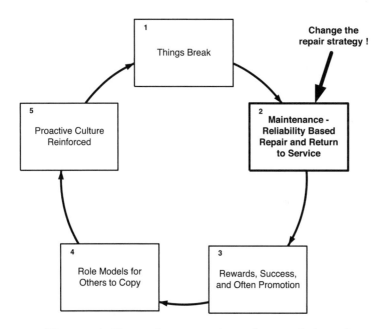

Figure 5-4   Example – proactive culture reinforced

the organization looking for someone to emulate. They are looking for someone who represents success and shows how things need to get done around here. So who do they copy? Not the preventive maintenance foreman (if one exists) because they are not the ones who get the rewards. Instead they copy our "fire fighter" and, in block 5, the existing culture is reinforced. Therefore, when things break the cycle is once again repeated.

The question that needs to be answered is how do you create new role models so that the organization will emulate a different behavioral style? You must change the organization's focus by breaking the reinforcing loop. Figure 5-4 shows that changing the second block (how the organization views and makes repairs) breaks the cycle. If you can alter how things are repaired, then rewards will be given to those who fit the new model (block 3). People will see a different definition of success and emulate it (block 4) and when things break in block 5 the organization will respond in a different manner. Therefore, role models are created, not born. You identify the behavior you want the organization to model, then place people who exhibit that behavior into jobs of importance

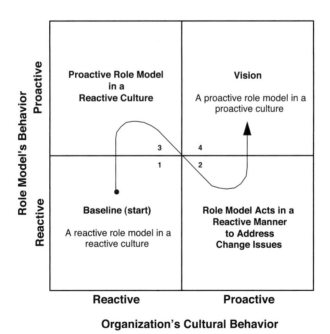

Figure 5-5 Cultural behavior vs. role model behavior

where they can visibly demonstrate the behavior you seek and, as a result, be copied.

An important point that needs to be recognized is that the site leadership must want to make this change. Although a new culture can spring up in one that has a different focus, it will not live a very long time. The predominant model and the supporting culture will extinguish it before it has a chance to thrive.

But what if the leadership does want to make the change? In this case, the existing role models will be given a different set of expectations. They will either comply or, over time, be replaced. The other alternative, one which is used more frequently than the former, is to bring in from outside the organization people who exhibit the new cultural behavior. They remove the old role models who can not adapt to the new way of working; they promote those who have adapted. The organization then sees these new managers as successful and, in the true definition of role models, emulates them.

5.7 How Role Models Can Change a Culture

Role models are a critical part of an organization's culture for several reasons. First, people need examples of how the organization expects them to behave in their jobs and within the larger context of the organization. Second, people want to see what success looks like so that they can model their behavior and achieve success in their own careers. Third, people also want to see the "good – bad examples" which depict clearly what the organization does not want to see in its employees. The first two reasons are positive and the third negative, but each has its place in the value of role models in the business

Roles models are needed because they are an essential and initial step in the change process. Figure 5-5 shows how role models fit and interact to help support cultural change. In this quad diagram, role models are portrayed on the y-axis as either reactive or proactive. The x-axis shows the organization's cultural behavior, also reactive or proactive. Note that role models can and do work in both roles at different times, dependent on the status of the change initiative. The line starting in block 1 and running through blocks 3, block 2 and ending in block 4 shows how role models introduce and bring about change in a company.

Block 1 represents the current condition or baseline that is encountered when a role model is brought into a company. Usually the existing role models have a reactive focus and, as a result, so does the culture.

Invariably the new role models recognize that change is needed; otherwise, they wouldn't have been hired in the first place. They next progress to block 3 (a proactive role model in a reactive culture) where they implement a change that supports their new model for doing work.

Initially a single initiative is all that is needed. We are not trying to "build Rome in one day." We are, however, trying to initiate long-lasting change and that must start slowly. For example, role models may mandate that a preventive maintenance program be established and not permit deviation from the schedule. Through their own behavior and through setting expectations for the organization, the role models provide guidance in how they expect work to be conducted. They do not pull the PM crews for reactive work, no matter what is happening in the plant. Furthermore, they praise those who have done the PM more then those who fight the fires.

Next we move into block 2 (a reactive role model in a new proactive culture). The culture is beginning to shift and become more proactive. At this time, however, the role models need to switch into a reactive mode and rapidly address those areas that are having difficulty. Small groups within the organization will not understand or act as expected, people will resist, and training will be required. Each one of these areas needs corrective action in a reactive manner if long-lasting success is to be the result.

Last, in block 4,because they have initiated a successful change in the organization, the role models can begin to introduce additional goals that support the new vision that they are trying to implement.

Following this path is not easy and takes concentrated efforts on the part of the role models. Remember that the organization is looking for examples that the role models are not serious so that it can return to the status quo and the existing personal comfort zones. See how long the new initiatives and role models will be in place if the PM crew is reassigned to reactive work even once.

Role models set the stage for everyone else; they clearly demonstrate every day what is expected of the organization. As an example, let us consider two reactions to a pump failure which has a severe effect on production. As you read each of these responses, think about the role that is being modeled by the manager.

Response #1 – The manager shows anger and frustration, and then asks how soon the equipment will be returned to service. The manager indicates that sooner is not soon enough and insists the equipment be worked on around the clock until the repair is complete.

Response #2 – The manager shows frustration, but then asks those involved to get the work completed as soon as possible with one exception. The manager wants the maintenance engineer to understand the mechanism of failure so that repairs can be made to prevent the failure from reoccurring. If this can not be accomplished during this repair, then a plan needs to be put in place to make the reliability-based repair at a future date when production can shutdown the equipment.

5.8 Can Consultants Be Role Models?

Consultants can be role models, but only in a limited sense. Part of what consultants bring to the process is strategic thinking unencumbered by the day-to-day activities. They can introduce new concepts that will support the company's vision and goals. They have a broad set of experiences that can add real value. They can also provide the training and coaching necessary to help get a major change initiative off of the ground.

During a consultant's engagement, in which they introduce proactive reliability-based maintenance processes, the consultants often take on the mantel of role model. One key reason for this is that they are the ones introducing the new model in a firm in which that model most likely does not yet exist. Therefore, it is no wonder that people emulate consultants. This is acceptable for the short term. However, consultants eventually finish their work and leave. Prior to this happening, company role models must be in place and be viewed by the organization as such. Otherwise the time, effort, and opportunity that the consultants provide will be lost simply because the company will not have role models in place to take up where the consultants left off. Without this internal role model, the organization will fill the void which the consultants leave by returning to the status quo. That is not what we want to happen.

I worked on one project that clearly shows what happens when a consultant / role model leaves and there is no one internally to fill the void. The effort involved the implementation of a new planning and scheduling process. The consultant we hired had a great deal of process change experience in this area and as a result drove the change. Initially the effort was successful because the role model for the process, the consultant, was there day-today helping us overcome the problems. We did not develop an internal role model to be his successor and when he left the program stopped. Quickly the organization reverted back to the former way of conducting this part of the work process.

The lesson to be learned is that you can have a consultant be your process change role model. However, you must develop internal resources to take the effort over when they leave. In fact, you should plan to take over before they leave so that in the closing days of their engagement they can provide coaching to the internal resources.

Rites and Rituals

6.1 Introduction to Rites and Rituals

Rites and rituals are a key component of organizational culture. When presented in various texts, these terms are often confusing because they do no easily relate to those in the middle levels of the organization; as a result, they often are not understood. The reader is left without a clear recognition of the critical role that rites and rituals play in an organization's culture and hence in an organization's ability to change that culture. This chapter will demystify these two terms, showing the part they play in successfully changing a culture from one that is reactive to one that is proactive and reliability focused.

As we discuss these two topics, remember that rites and rituals are invisible to those involved in the day-to-day work. They are so ingrained in how we work that we simply consider them as the correct way things are done. This causes a problem on two levels. First, by not being able to see the forest for the trees, rites and rituals are difficult to identify clearly and, therefore, even more difficult to change. Second, because they are so ingrained, when change does actually happen we often feel like our world has been torn away. Being this far out of one's comfort zone leads to frustration, anxiety, and serious levels of resistance.

There is a story about a wife who, when making steak for her family, always cut off a small piece from the end before cooking. When asked why, her answer was that this was how her mother taught her to cook steak. Not satisfied with this answer, and not wanting to waste good steak, her husband asked his mother-in-law the same question. Her reply was that this was how her own mother had taught her how to make a steak. Still not satisfied, the husband asked the grandmother the same question. Her answer, which surprised the entire family, was that there was nothing wrong with the end of the steak. However, when she first started cooking, she didn't have a pan large enough for a normal size steak, so she had to cut off the end. This is a ritual at work.

In years past, I worked for a company that held quarterly performance reviews. Once per quarter, the senior corporate staff would make departmental presentations that addressed goals and accomplishments. In my opinion at the time, these presentations wasted hours of potential productive time for the employees, listening to presentation material that was already known by everyone in attendance. What I didn't understand at the time, that will be discussed later in this chapter, is that what was taking place was a very important rite for the company.

My goal in this chapter is to help you understand your hidden rituals and the rites that they create. Then, when necessary, you can identify and modify them to support a change to a reliability-focused work culture.

6.2 What Is a Ritual?

A ritual is a rule or set of rules that guide our day-to-day work behavior. Rituals are taken for granted because they are an integral part of what our jobs are and how we do them. As they are repeated daily, rituals become an accepted part of how business is conducted; over time, they become invisible to those who follow them. Yet they are extremely important not only because they define what we do and how we do it, but also because they represent our culture and the value system in place in our plant.

Furthermore, rituals are taught to new employees so that they understand "how work gets done around here." Rituals guide how people communicate and interact. Because they are so ingrained in our work, an outsider might say they were blindly followed – often even if they made little or no sense. In addition, they are often fiercely defended simply because "that is how things are done." This viewpoint explains to some degree why new programs or processes that conflict with the plant rituals encounter such strong resistance when they are implemented.

Every one of us has had this experience. We are assigned a new job. The first thing we are given on our first day of work is training in how the work is conducted — the rituals of the job or department and, what is more important, the culture in which we now reside. As a new employee, this training is highly important because we are being told not only how to act but also what is needed to be successful.

Many years ago, I received my first supervisory opportunity as a foreman at my plant. Before the foreman who I was replacing left for another area, we spent an entire week together. I learned how work was assigned to the workforce, how to interact with production, how to order materials, and many other tasks.

At the time, our plant was totally reactive in the way we conducted maintenance. When things broke down, our most important task was to repair them as quickly as possible and return them to service. Still I was surprised when after lunch on Friday the entire crew was not assigned additional work but stayed at their staging area. I questioned the foreman I was replacing and was informed that they were waiting for things to break so that they could rapidly respond to the problem, make the needed repair, and avoid weekend overtime. Being naïve I asked why they couldn't be assigned jobs that could easily be interrupted. That way, we could get some work accomplished while at the same time be available to respond to plant emergencies. I was told in no uncertain terms not to "rock the boat" because "this was how things are done around here." This was the ritual followed by each foreman. The culture was not about to let me change it!

In our context, therefore, a ritual is an invisible day-to-day work practice that is accepted as how work is performed within the organization's culture. The ritual provides everyone with a foundation for how work is handled. Processes outside of the accepted rituals are considered alien. The organization will feel extreme discomfort when new rituals outside of the accepted norm are introduced, even though it may not know exactly why. When I suggested an alternate solution to waiting for things to break, I was reprimanded even though the outcome would have been the same — we could have still responded to production's emergency needs.

6.3 What Is a Rite?

A rite is a company ceremony or event that reinforces our rituals. In a sense, they provide a stage for those involved to dramatize the culture and organizational values to those in attendance. Rituals and rites go hand in hand because without accepted rituals, rites do not exist.

Rites can cover a large spectrum of an organization's events. They include performance reviews, training, conferences, service awards, and departmental and group meetings, all the way down to a pat on the back for a job well done.

One company I worked for years ago had its headquarters in California, although our plant was on the east coast. We had no regular visits from senior management except for the quarterly review. This meeting was held every quarter when the senior management team came to the plant. Senior management arrived in the evening, then spent the next morning attending a large plant meeting at which each department presented its accomplishments from the previous quarter and its goals for the next quarter.

I was involved in many of these efforts and often questioned why this was necessary. The senior managers received weekly updates so that everything presented was known beforehand. The plant personnel in general had this same level of knowledge, having been involved with the majority of the work on a daily basis. In addition, there were at least fifty people in the audience representing all of the departments. This amount equated to expending at least 25 work days attending the meeting, not even including the preparation time.

On the surface, this effort may appear as a waste of time — it did to me. However, in the context of our discussion of cultural rites, this event had a specific purpose. It was a stage to dramatize for the middle and upper management the culture and the values of the company. The topics that were discussed were the department's goals and the progress being made towards their attainment. As these presentations were delivered, the senior staff asked questions and made comments. Some praised the work that had been done; others did not. From what was said, it became extremely clear where the work focus would be placed. In addition, the corporate staff made a presentation regarding the state of the business at the end. This rite specifically reinforced the company culture, the value system and the rituals of the plant.

Let us look at a simpler example. Consider the foreman who kept his crew in their staging area on Friday afternoon waiting to respond to the emergency needs of production. Several rites were associated with this single ritual. The first of the rites is the "pat on the back." When production called, maintenance was available to make the quick fix. If successful, the foreman would get a pat on the back for a job well done — a rite positively reinforcing a plant ritual. This sort of success would eventually translate into another rite — a positive performance review, better salary, and a chance for promotion.

Conversely, if the ritual was not followed, the associated rite would have a severe negative connotation. In this case, production would complain about the foreman's performance, resulting in other potential problems for the foreman who was out of compliance. My idea of having the crews work on interruptible jobs on Friday afternoon not only violated a maintenance ritual, but also seriously threatened an established set of rites for the foremen — the pat on the back and others of more significance.

Let us look at one more example at a personal level – the service award. Companies like employee longevity. It provides an experience base on which to build the business, it provides people who can train and mentor new employees, and it reduces the amount of lost effort as new employees learn how to conduct the work. The ritual is simply productively working on a day-to-day basis. Many companies award those who have remained employed and provided years of company service. As one advances in years of service, these gifts increase in value. In other companies, the service award is a dinner with the senior managers. There are other types of service awards, but they all represent a company rite – the attempt to recognize employees with extended service and visibly show others in the organization the value that is placed on these employees.

Just as with rituals, rites exist throughout our corporations. They are easier to identify because they are significant events – the performance review, the service award, or even the production manager telling the foremen that they did a good job on Friday's emergency breakdown. These rites, or methods that reinforce the rituals, are important because they perpetuate the ritualistic behavior desired.

6.4 How Rites and Rituals Are Tied Together

Rituals create the need for rites, which provide the way that the culture reinforces the rituals that people live by. In turn, the rites reinforce the rituals by providing feedback that what was done was the correct action for the culture.

These two events create a loop. Rituals are performed and then the behavior is reinforced by the rite. Those involved then repeat the ritual to continue getting the reinforcement that they desire. This loop is represented in Figure 6-1.

To describe this process, let us use the example of the work crew waiting for the Friday breakdown. In this case, the ritual consists of the crew waiting for the plant emergency, which by experience always seems to happen on Friday. When it occurs, they get the call and within minutes are on the job. Production is comfortable because its problem

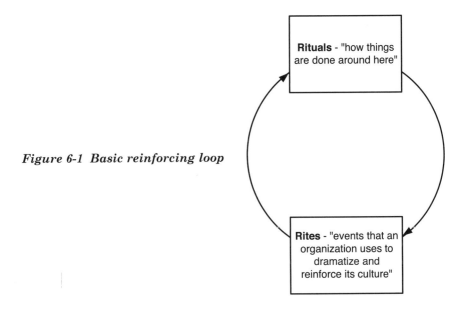

Figure 6-1 Basic reinforcing loop

is being rapidly addressed. Most often the work is completed and production is restored. At this point, the ritual has been completed. The next step is that the production manager tells the foreman "great job." Because this individual is often in a position of power, the foreman feels good about the effort expended. It is rapid and positive feedback, an ego builder for the foreman and crew. Because these events occur often, the production manager will probably praise the foremen to his manager, which then may result in a raise or promotion. Thus the rite performed by the production manager has reinforced the behavior of the foreman. Praise, promotion, and possibly a very good raise are rewards for performing the ritual. Conversely, if the ritual is not performed, negative feedback, and other more serious career threatening events may occur. Recognize from this discussion how hard it is to change a ritual. If you wish to be successful at change, you must change the rites that reinforce the ritual as well.

6.5 Determining a Rite from a Ritual

As we have discussed, rituals are performed and then reinforced by rites so that they are repeated. Although this loop is hard to break, we will need to do so. Our ultimate goal is to initiate a reliability work culture. Therefore, the actual reactive rites and rituals, as well as the reinforcing loops, need to be replaced.

We first need to be able to determine the rites and the rituals within our company. At times, we can review an activity and clearly recognize it as a rite or ritual; at other times, this distinction is not completely clear.

Four characteristics will help us differentiate rituals from rites. By reviewing events in the context of these four characteristics, you should be able to determine if the event is a rite or a ritual.

1. The Frequency of the Event within the Work Process

Events that have a high frequency, such as unplanned equipment failure and the required emergency response by maintenance, are usually an integral part of the work process. These are most often rituals defining how the work is to be done. Because rites results from the performance of rituals, they occur on a much more limited frequency, such as quarterly performance review.

2. The Rapidity of Cultural Reaction

When you change or remove a ritual, you immediately alter how the work is being done. As we shall see, this change elicits a rapid response from the culture being affected. Rites, on the other hand, are a result of, but not a direct part of the work process. Therefore, the organization is not affected as quickly and the reaction is slower to manifest itself.

3. The Depth of the Reaction

Changing rituals often provokes a strong response from those affected because you are directly impacting how they perform work. Such a reaction results from altering the status quo and causing organization discomfort. The reaction is to attempt to restore the norm. Rites do not carry the same depth of reaction. Because rites are reinforcing mechanisms, you are not directly impacting the work and those who perform it. For example, if we eliminated the Friday emergency crew we would get a strong reaction to the change in the ritual. However, if we eliminated the praise from production for the effort — the rite — the foreman would not be as satisfied, but the work process would be unaltered.

4. Part of the Work Process

Rituals are typically a direct part of the reliability and maintenance work process such as implementing a PM program. Their supporting rites, on the other hand, are usually the result of properly executing the rituals — praise by management for reducing failures as a result of the PM program. The rites provide the reinforcement of the rituals. Differentiating rituals from rites is simplified by examining the event to determine if it is a direct part of the process or one that reinforces it.

A few examples to describe these reactions may help.

Event #1
Maintenance management determines that crews sitting in their staging areas on Friday afternoon waiting for things to break is a poor use of the maintenance workforce. As a result, work is assigned that can easily be interrupted so that crews can be diverted to the emergency as needed.

Type: Ritual
Level of Reaction: Extreme
Frequency: Short duration – weekly
Part of the Work Process: Yes

Rapidity of Reaction: As soon as this change was announced, the production manager called his maintenance counterpart to complain. He cited examples where these crews saved his operation. Even though the maintenance manager explained that service would not be any different than before, the production manager threatened to go to the plant manager with his complaint.

Event #2
Management recognizes the amount of time expended for the quarterly review process and decides to have only department managers present. They believe that these managers can communicate the relevant information to their organizations.

Type: Rite
Level of Reaction: Low / mild
Frequency: Long term (once per quarter)
Part of the Work Process: No

Rapidity of Reaction: Slow or none. Some people would object, but there would be limited reaction over time. In fact, many would welcome having one less meeting to attend.

Event #3

The maintenance department decides that there should be a dedicated preventive maintenance (PM) crew for rotating equipment. However, there is no approval to hire additional resources. Because the goal is to perform high quality PM and troubleshoot at the same time, the top mechanics are assigned to this crew. This change results in fewer resources to perform breakdown maintenance.

Type: Ritual
Level of Reaction: Extreme
Frequency: Short duration – this work is conducted daily
Part of the Work Process: Yes

Rapidity of Reaction: Immediate. As soon as production learned of this change, there were extremely strong protests from the area production supervisors. They believed that the level and rapidity of repair would suffer. They concluded that their operation would be at risk with fewer resources working on their equipment.

Event #4

In order to save money, the company decided to eliminate the service award dinner and handle recognition for service at the individual's supervisory level.

Type: Rite
Level of Reaction: Neutral to none
Frequency: Long term – once per quarter or lesser frequency.
Part of the Work Process: No

Rapidity of Reaction: Moderate to none. Some people would be mildly upset because they missed out on a free meal, but in general eliminating the dinner would go unnoticed.

These examples do not always follow the pattern. Suppose that in the fourth example the employees placed great value on the service award rite. If this were the case, then its elimination would have elicited a strong and rapid response once the change was known. There is no easy way to differentiate rites and rituals unless you know something about the culture.

Figure 6-2 helps us to visually recognize this differentiation. This chart compares the four key characteristics of rites and to help you determine which type you are addressing in your analysis.

Characteristic	Rituals	Rites
Frequency	Short	Long
Level of Reaction	Extreme	Mild
Rapidity of Reaction	Fast	Slow
Direct Part of the Work Process	Yes	No

Figure 6-2 Identifying rites from rituals

6.6 Values, Behavior and the Rites and Rituals They Create

Rites and rituals are elements of an organization's culture, but they do not stand alone. They do not exist in a vacuum. They are linked to organizational values and the behaviors that are created by these values.

The reinforcing loop in Figure 6-3 shows this relationship. We will use a reactive work culture as the example to make our case. The plant environment, shown in Block 1, is one in which things continually break down. The ritual of the quick fix then kicks in (Block 2) and maintenance rapidly responds to correct the problem. Over time, without any change in the plant's "break it – fix it" value system; these responses manifest themselves as rituals. As maintenance continues to step in and save the day, this level and type of performance is reinforced by the organization through rites (Block 3) created for this purpose. These appear as the pat on the back and the "atta boy," or even more significantly as raises or promotion. These rites reinforce the rituals, supporting a continued focus on reactive maintenance so that when things break, the process is repeated (Block 4).

This process is very important. It serves to explain the difficulty of making a change to a new way of working when the culture is firmly anchored some place else.

Did you ever wonder why good ideas resulting from training programs or conferences you attended failed to take hold? They left you with good memories of the program, but little or no benefit to your

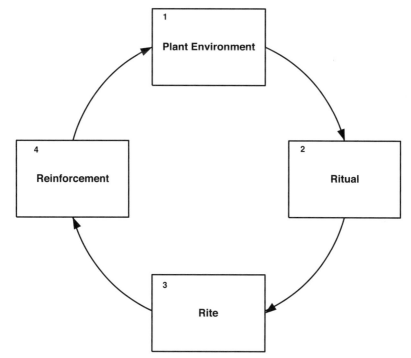

Figure 6-3 Reinforcing loop with rites and rituals

organization. The answer can be seen in Figure 6-3's reinforcing loop.
What you brought back were new concepts about your work. You tried
to introduce them as new rituals. Immediately they came into conflict
with the collective organizational rituals (Block 2) and rites (Block 3)
which set them up for failure. Your ideas were not "how it is done
around here." As a result, they were never adopted.

As hard as it appears to be to bring successful change to an organization, the level of difficulty does not mean that new ideas can not succeed. It just means that most often they can not be brought in without
a lot of hard work. Furthermore, for success to be achieved, the new
ideas must be brought in at the ritual and rite level (Blocks 2 and 3). In
this manner, you can change the essential way things are done (Block 2)
and the mechanisms that reinforce them – the rites (Block 3). Over time
and with a great deal of work, the rituals will change and the culture
will develop new rites to reinforce them – Block 4

6.7 From Reactive to Proactive Maintenance

So far in this chapter, we have discussed rites and rituals, and how hard they are to change within an organization's culture. We have also described how they do not exist in a vacuum, but are a part of the culture. But changing a culture is not impossible. A great many companies that were formerly focused on reactive maintenance have changed and are now recognized for their pacesetter reliability-focused work cultures.

In changing from reactive rituals and rites to ones that are proactive and reliability-focused, we need to make several key assumptions. We know that rituals and rites are part of a larger whole. Therefore, we must make assumptions with respect to organizational values (Chapter 4) and role models (Chapter 5). We first assume that the company or plant's senior managers have adopted a reliability-focused value system in which it is expected that things do not break down. Here, reliability is paramount because these managers recognize it to be a critical component of production. Second, we assume that our role models (Chapter 5) are those who visibly and actively support the reliability focus. These role models also work diligently to eliminate the prior practices associated with the reactive mode of operation.

If values and role models are not in place, actively promoting the concept of reliability, then altering the rituals and rites will be impossible; the switch to a proactive work culture will never happen.

The plant in our example has a new value system where reliability is characterized as a vital component for the success of the business. In order to promote these values, role models at the senior managerial level have been brought in from outside the company. They believe in these values and have been successful at other companies promoting a reliability-based work culture. However, in our plant, reactive maintenance is still the norm and is still being reinforced by middle management.

The new managers must change the rituals and rites to ones that support the new value system. This will not be easy and will elicit resistance.

The question then is how the new management team makes the change. Some believe that you need to negatively reinforce the behaviors you want to eliminate and positively reinforce those you want to promote. In other words, punish reactive work efforts and praise proactive work efforts. I don't believe that this is a sound approach because some will use the negative reinforcement as a rallying point for resistance. They will use the negative reinforcement to show others how the new management "are destroying what we have created." This approach is further complicated if everyone is not trained in the new approach. If employees who have not been trained in new methods receive negative reinforcement for working the old way, the organization will be lost. Being out of your comfort zone is bad enough, but not knowing how to restore a level of comfort is highly stressful.

A better approach is to positively reinforce the new way of working and not reinforce the old way either positively or negatively. With proper training in the new way of working, the organization will be able to move forward. As employees perform in a proactive manner, they will get reinforcement. Over time, this will evolve into rituals and the associated reinforcing rites. As for the former reactive work process, without any form of reinforcement, it will disappear.

In the new reliability-based work culture that we are trying to create, our value system and the management team in place are driving a process in which things don't break. For the reactive culture of the plant, this is a very different approach to the work when compared to how it is currently performed.

How do you, as part of the management team, make the change? You begin by altering the rituals that are part of the reactive work environment. You eliminate the crew sitting in their staging area Friday afternoon waiting to respond to an equipment failure. You institute crews whose task it is to perform preventive maintenance and you do not interrupt their work – no matter what. You break up the "good old boy" relationships between the foreman and the production people who are always calling with the emergency of the day. In the end, you redesign the work process so that it is entirely focused on proactive maintenance processes.

In addition you measure the success or failure and then visibly reward those who work towards making it a success. However, you do not punish those who still cling to the old reactive rituals. Instead you develop job expectations for these individuals that are proactively focused, rewarding them when they comply. If they don't, you continue to coach them until it becomes obvious that they have no plans to ever change. At that point, you remove them from the process. This isn't negative reinforcement, this is corrective action.

Very soon it will become clear to the organization that there is a certain new set of rituals to follow. Success in instituting these new rituals will give way to rites that will support them. These will not appear by

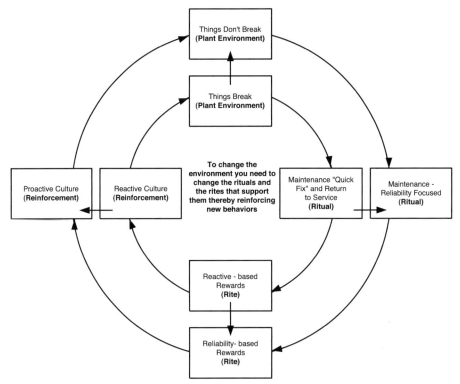

Figure 6-4 Example of changing the reinforcing loop

magic. As a manager trying to implement a new order, you will have to create these rituals and rites.

This process is depicted in Figure 6-4. The inner circle portrays the reactive process – things break, the rituals in existence support reactive repair, the rituals are accompanied by rites that reinforce them so that when things break the process is repeated. The outer circle shows what you are trying to achieve. As we discussed, new rituals are put into place that support a proactive reliability-focused culture. Along with these rituals, rites are instituted that reinforce the new processes. As a result and over time, the new processes are reinforced so that "things never break." Along the way, those who cling to the old way of doing things receive no reinforcement for their behavior. Instead they have goals set that attempt to refocus them on the new order of things. If they don't comply, they are not punished, but rather receive continued coaching. Over time, these individuals will change or voluntarily leave.

The Cultural Infrastructure

7.1 The Cultural Infrastructure Identified

What do all of these scenarios have in common?

You attend a meeting where an important proposal is presented and approved. You later learn that all of the key decision makers had already reviewed and approved the proposal. The meeting was just the final group approval, not a decision point within the approval process.

You are forced to listen to war stories about the plant and its personnel at almost every meeting you attend. You believe that the time spent repeating the same stories detracts from the meeting's value. However, when you examine the content of the stories, you notice a close connection to what you know of the organization's values.

You notice that several employees who have been with the company for many years are always consulted by others seeking direction for their work. When you look into the reasons, you realize that these senior employees are explaining how to get work accomplished successfully within the company.

You are aware that if you want your boss to know a specific sensitive piece of information, but you don't want to bring it up directly, all that you need to do is to tell your boss's administrative assistant. When you do this, the boss always seems to find out.

You remember with some pain the time you criticized one of the senior managers behind her back about what you felt was an incorrect decision. Later you were brought to task by your manager for your comments. Because you did not make the statement directly to the senior manager, someone else told her what you had said.

You attend a group meeting with people from another department. After the meeting, they inform you that they had no idea of what was said because everyone was speaking in acronyms. They said they felt as if they were in a foreign country where they didn't speak the language.

Your office is being renovated. Walls are being moved and new office spaces are being created. You tell your manager that the floor plan has a lot of unused space and that, in order to eliminate it, some of the offices could be enlarged. You are informed that such changes are not possible because the company has a corporate policy that assigns office space based on position within the company.

These seven examples all have something in common. They are all real examples of the cultural infrastructure at work. Throughout the first six chapters of this book, I have described aspects of the organization's formal culture. However, another level of culture is equally important, although hidden – the cultural infrastructure. Just as the value system, rites and rituals, and role models describe the culture of the organization, other attributes describe the cultural infrastructure.

7.2 Cultural Infrastructure – Definition and Component Parts

The various components of an organization can be represented as blocks in a flow diagram. Although the blocks each represent a function within the company, they can't stand alone. They need the connecting lines that tie them together, providing a linkage for all of the individual parts. This linkage is the cultural infrastructure.

For our discussion, we will focus on people and communications as the key elements of the cultural infrastructure. These components are the glue that binds the organizational culture together and promotes sustainability of the firm. Thus, our definition of cultural infrastructure is as follows:

> Cultural infrastructure is the hidden hierarchy of people and communication processes that binds the organization to the culture and sustains it over time.

Although by definition the cultural infrastructure works towards sustaining the existing culture, it can also be used to promote cultural change.

In their book *Corporate Cultures,* Deal and Kennedy describe the various aspects of the cultural infrastructure. I have modified their list and added two more. Whereas Deal and Kennedy have taken a generic approach, my focus is on how the cultural infrastructure affects plant reliability. The aspects of the cultural infrastructure that I will discuss are as follows:

Story Tellers – promoting the culture through war stories

Keepers of the Faith – mentors and protectors of the culture

Whisperers – passers of information behind the scenes

Gossips – the hidden day-to-day communication system

Spies – passers of sensitive information to those who may or may not need to know

Symbols – mechanisms for conveying what and who is important

Language – terminology that describes what is done and often how

7.3 Story Tellers

Every company has story tellers. These are the people who at a meeting launch into a story of how some important thing was accomplished and, often very specifically, how it was done. All of these stories have something in common. They are not neutral – they are stores of success or failure and always related to specific events in the company's history as well as the specific people who performed them. The message that these story tellers are conveying ties into our discussions in Chapters 4, 5, and 6. The stories show organizational values, the rituals and rites that were employed, and the role models who accomplished them. The stories try to perpetuate the organization's culture.

I have worked with a consultant firm whose members (including the owners) tell stories all of the time. Many have funny components, but in each of them the message is clear – "we have a good time doing what we do, but we always get the job done no matter what it takes." These stories serve several ends. First they link the members together with a shared history. You feel part of the process even if you were not an active part of the story. Second they convey the value system of the owners. Because it is a small company, satisfied clients are the foundation of the business; the firm has built a close, long-lasting relationship with each of them. Third it is a way to show new members what company and personal success mean and what is required to achieve them.

But what if you work in a company whose stores convey the wrong message? What if your company's storytellers are preaching the old way of doing business (e.g., reactive maintenance), but you and your management team are trying to change to a reliability-focused process. The problem is that the storytellers are reinforcing something that you want to change. By the very nature of their stories, they are blocking progress to a new and better work culture.

The solution to this is to change the stories to reflect what you are trying to achieve. You need to disconnect the stories from the past and orient them to the future. There are two ways to do this. First, you need to be present on occasions when a story is told about how the old way of working was the best, then refute it. The second solution is to change the stores that the story tellers tell. This solution is clearly more likely than the former.

To accomplish it, you must first identify your organization's story tellers. They are not hard to find. They are the people who are always relating war stories about the old way of working to anyone who is willing (or even not willing) to listen. They are found at any level of the organization and seem to be present at most, if not all, meetings, especially when the meeting is focused on change.

Once identified, you need them to become a key part of the change initiative. Depending on their role within the organization, this task may or may not be an easy one. Nevertheless, it must be done. You need them to embrace the new order so that their stories reinforce the new instead of reinforcing the old and blocking progress.

Visualize for a moment a meeting where you are trying to get the organization to change from a reactive maintenance process to one that focuses on preventive maintenance strategies. Prior to enlisting the story tellers in the new process, they would have provided the organization with tales of how maintenance quickly reacted to production's needs, saving the day. In their new role, their story would be far different. It would relate how the PM program would have eliminated the failure from ever happening in the first place.

Granted this task is not going to be easy, but short of being at every meeting and personally stopping the stories that reinforce the old way of working, it must be done.

7.4 Keepers of the Faith

Keepers of the Faith are an entirely different group. They are the coaches and mentors of the organization. We go to them for guidance when we have a question about how to proceed. They may be the firm's managers or they simply may be individuals who have been with the company for a long time and know how things are done. What is important is that when we solicit their council, they provide the basics of how things get done within the context of the organization's culture.

Several years ago I was involved in a project designed to replace the various computerized maintenance management systems in our plants with a single system. Along the way I discovered that to be successful, you needed first to change the work process, then to support it with the new computer system. However, my company believed that the work process would improve if we simply installed the new software.

My presentation to senior management focused on what I knew to be true – work process changes followed by software implementation. The first time the senior management team heard my approach was at the meeting when I asked for funding for the project. Surprisingly to me, I was thrown out of the meeting; I never got to finish my presentation nor make my case for change.

My manager at the time (a true keeper of the faith) took me aside. He explained that although my idea was sound, my approach was wrong. The company did not make decisions at meetings. Instead, decisions were made in advance through one-to-one contact with each of the staff. In this way, they had time to ask questions in a non-political atmosphere and decide in advance if they were going to lend their support to the effort.

What my manager was explaining was how the culture worked if one wanted to succeed. This was based on his personal experience with the organization. In response, I personally met with each of the staff and presented my ideas. It took a lot longer than doing it one time at a meeting, but I was able to gain acceptance. Amazingly, when I presented the concept at the next senior staff meeting, what had been rejected out of hand on the previous occasion was unanimously accepted. The culture did not allow for major decisions to be handled at staff meetings without time for each member to individually consider the proposal. This story worked out well because the company was already proactively focused. My boss simply showed me how to accomplish my project successfully within that context.

But what if the Keeper of the Faith is promoting a culture that is not proactive? What if the organization sees no advantage to changing from a reactive focus to a reliability focus? In this case, the Keeper of the Faith is a roadblock to change. Suppose that after my failure in the first presentation, I was taken aside and reprimanded for suggesting that we change to a more proactive work process. I could have been told that the process was fine as is and that my job was to make the new software work within it. In this context, any Keepers of the Faith would actually have being working as a road block to the change process. The question then is: What can be done to remove this rather subtle roadblock to progress?

The answer is similar to that taken for the Storytellers. They need to be enlisted into the ranks of those who believe that a new and proactive approach to maintenance is what is needed. Convincing them may be very difficult because, in their role, they act as the caretakers of the old way of doing things. Nevertheless, if you wish to get behind-the-scenes support for proactive process change, you need to change the beliefs of these individuals. In this manner, when it comes to explaining "how things are done are here," they will relate the new vs. the old way.

7.5 Whisperers

Whisperers are those members of the organization who have the ability to whisper in the boss's ear, hence their name. Whispering is their source of power because they are not typically leaders of the organization. However, they play an important role within the cultural infrastructure by conveying information upwards that may not otherwise be communicated within normal channels. Whisperers are most often trusted subordinates of the management team. They have achieved this position usually from long years of trusted service. Therefore, managers feel secure in the information that they provide. There are even circumstances where the managers rely on this information as a pulse of the organization – something they would not be able to obtain through other methods.

There are numerous examples of whisperers at work. For instance, how often in casual conversation do you mention something to your managers' administrative assistant only to hear about it later from the managers themselves. You know that there is no other way that the managers could have obtained this information except from the person you told – one of the organization's whisperers.

These people can have a positive or a negative effect on cultural change, depending on what they pass along and how it is done.

The communication process has several steps between what you want to communicate and how it is received. These steps include: determining what you want to communicate, establishing the means of communication (e.g., verbal, written, visual), transmitting the message to the receiver, and finally the reception and interpretation of your message. This is a rather complex process.

The place where whisperers fit into this process is in the transmission of the message. You may not even want to send the message at all. In this case, they are passing along information without permission. The result may be that the managers get information that you never wanted them to have. If this is the case, then you have learned a valuable lesson. Do not provide information to a whisperer that you don't want passed along.

The second communication issue with whisperers arises in that the message is not always delivered as sent because of the filtering process that takes place. This filtering can be minor in nature and the message will be delivered fairly intact. However, the filtering can also be major, resulting in a delivery of a message that bears little resemblance to the original message or has an entirely different context.

Suppose you tell a whisperer something that you want the boss to know, but do not want to communicate directly. If the message is passed along with minimal filtering, then is no harm is done. Your attempt to communicate indirectly is successful. Now suppose you make a remark, just in conversation, about a decision made by your manager. Yet the whisperer changes the context, delivering the message as if you are personally questioning your manager's authority. Or suppose that you and a co-worker are having an angry disagreement about a decision your manager has made. You are supportive of the decision, whereas your co-worker is strongly opposed. However, the whisperer inadvertently reverses your positions, telling your manager that you are opposed to the decision and your co-worker is supportive. In both of these cases, you may be in line for some serious trouble.

How can whisperers help support cultural change? As with the other two components of the cultural infrastructure (Storytellers and Keepers of the Faith) you must identify who the whisperers are and then engage them in the cultural change process. You need their support because they have something that will help support change – the boss's ear. By engaging them in the process, you can get their help in communicating behind-the-scenes information that may not otherwise be communicated along normal channels. You can even have a degree of confidence that the filtering (if any) will be supportive of change.

7.6 Gossips

Gossips are like whisperers, except they communicate to a different audience. Whereas the whisperers communicate upwards, the gossips communicate throughout the entire organization. They spread stories, rumors, and other sorts of information. They are part of the plant's informal communication system.

I was once called into my boss's office and told of a pending departmental restructuring. I was also told not to say a word about the changes until they were announced by the company. Being a loyal employee, I made up my mind to say nothing as requested. Within the next day and prior to any announcement, I had several of my employees as well as others in different departments tell me exactly what was to take place. If everyone had kept quiet, this information could not have been passed throughout the company. However, people didn't keep quiet. The gossips spread the information much faster than any company communication ever could have achieved.

Gossips exist everywhere and at every level of the organization. They can spread rumors that paralyze a company with negative information such as a pending layoff or plant shutdown. They can create false hope when none exists by passing false information about areas such as earnings reports or wage increases. Or they can reveal positive information that preempts a company's strategy such as acquisition of another company or release of a new product. Even though this information is positive, a formal release by the company, not the gossip, may have been a far better approach with accurate communication.

Unlike whisperers, gossips provide no benefit in support of organizational change. Although the whisperers can be enlisted to provide behind-the-scenes upward communication, gossips can not.

There are several reasons why this is the case. First, gossips derive pleasure and a sense of power by spreading information. If the information happens to be accurate, so much the better for those on the receiving end. However, the information does not have to have any degree of accuracy for the gossips to achieve the recognition they seek. Consequently, gossips not only filter information, as do the whisperers, but they also embellish it to suit the level and type of recognition they seek.

The second problem with gossips is that there are a great many of them. As a result, the information they pass will be distorted in many and different ways, making any use of this communication channel worthless. Remember when you played "whispering down the lane?" By the time the information got from the first person to the other end of the line, it was so distorted that the difference between the original message and what came out the other end was comical.

Gossips can not be eliminated. Recognizing that the communication value of gossips is often distorted, negative in its content, and often disruptive, what can be done to disrupt or negate the value of the gossip?

I believe the answer is rather simple. You simply don't give them anything to gossip about by closely controlling the flow of information – a difficult but not impossible task. Or you take a different tack and make the gossip's information worthless. This second approach can be accomplished by utilizing the Goal Achievement Model to lay out a long-term plan. Then provide a continuous stream of information to the organization related to the plan's progress. Because gossips thrive on spreading information to fill a communication void, remove the void. Provide your own real and timely information about you're the change effort. This will leave the gossips little to gossip about.

7.7 Spies

A spy is a person who obtains sensitive information by various and often devious methods, then passes it to people or groups who use it to gain advantage over their rivals or adversaries. In business, spies typically pass throughout the organization information that people do not necessarily want to be communicated. For example, you and your team fail to complete a portion of your project on time. Through a great deal of hard work, you are able to correct this problem and get the project back on track. However, you have a spy in your midst who immediately tells your manager of your team's failure.

Spies exist only if there are managers, other rival teams, or departments willing to accept to information and use it to their advantage. If we are trying to create a proactive reliability-focused work culture, and there are those in the organization resisting this change, then spies will have value. They will be the ones who provide these resisters with information that they can use to undermine the effort. They will be the ones to quickly tell of any failures so that those trying to block the change can exploit this information.

On the other hand, if we have managerial alignment with everyone focused on achieving the reliability outcome, the spies will cease to have any value for the organization and will most likely cease to exist. Without someone seeking the information that the spies have acquired, there is no need for the spies. They can not exist in a vacuum.

Unfortunately, creating an organization without spies and those who seek the information that they possess is virtually impossible. A solution is to negate their credibility or, as with the gossips, provide information and continuous communication so that nothing is hidden.

We usually know who the spies are in the organization. However, little can be done because the information that they provide is often to managers who will protect them. Therefore, to negate their credibility, you need to get those on the receiving end to doubt the information's validity. This is done through open and honest communication of all of your work – success and failure alike to the recipients of the spy's information. This communication will eliminate any information that the spy can pass that the manager doesn't already know.

A second alternative, although more risky, is to pass disinformation to the spies and the correct information to the manager. Consequently, when the spies report, the information they provide is wrong; their credibility is reduced and eventually eliminated.

If eliminating the credibility of the spy is not a solution you can adopt, you may want to take the same approach as you have for the gossips. By placing your work within the context of the Goal Achievement Model and providing continual updates, both positive and negative, you leave nothing for the spy.

7.8 Symbols

Symbols are that part of the culture that tells us who we are as well as how important the organization thinks we are in the scheme of things. Examples include office size and location, special privileges like a company car and the right to eat in the executive cafeteria, all the way down to different emblems or hardhat colors that depict which department (or sub-culture) we belong to. These symbols clearly state who we are without ever having to say a word.

Suppose you are walking down a hallway in your corporate office. If your culture defines people's importance by their office size and location, you will immediately know who the senior managers are by the fact that they sit in the corner offices. Next will be their direct reports, who will have window offices. They are followed by those lower in the organizational structure who will be sitting in the interior offices, often partitioned off from their co-workers as opposed to having a floor-to-ceiling office.

Other organizations are quite different. In these firms, people all have the same size offices and there is little discrimination as to what job permits the window offices. The message sent to the organization in these firms is one of equality and openness.

Symbols can also indicate the level of importance that a firm places on reliability. Take two examples of rotating equipment vibration programs and the people who populate them. In the first, the team does not have the latest technology and their offices are in a small section of the machine shop. In the second, the team has access to new equipment and the latest technology. Their offices are well maintained and situated close to the supervisor's office. Think about what the symbols say to the organization about the importance of the rotating equipment reliability program. The organization that considers this program an important part of their business is obvious.

One way to show the importance placed on a change initiative within the cultural infrastructure is to provide symbols to the change team that you want to differentiate in a positive manner. These symbols should indicate that support to anyone who takes a moment to look. The opposite holds true if you want to indicate that a specific activity or change effort holds little value.

7.9 Language

Part of the cultural infrastructure is composed of the language we use in our day-to-day communication with one another. This language is usually loaded with acronyms – initials that mean something to those who are part of the process being discussed. Understanding these acronyms clearly identifies who is part of the specific sub-culture. Those who are not knowledgeable in the sub-culture's language often do not understanding what is being said.

Each department has their own set of acronyms. For example the Reliability Group may refer to PM (preventive maintenance), PdM (preventive maintenance), RCM (reliability-centered maintenance), MTBF (mean time between failure), and other terms that describe their phase of the business. If you are part of the group that uses this terminology, you don't even have to take a second to think about what these terms mean. If you are not part of this group, you will not fully understand their meaning s in the context of the conversation, even if you recognize the term. As a result, you will feel left out.

Language can cause a split between those who are a part of the sub-culture and those who are not. Efforts that are designed to change the culture from reactive to reliability-focused require that the entire organization work together as a team. Therefore, a split caused by language is important to avoid. This goal can be accomplished by either eliminating the divisive language barriers or by educating everyone so that even when specific language is used, it is understood.

Take for example the preventive maintenance (PM) program. When they hear PM, production immediately thinks they will need to take equipment off line so that maintenance can perform tasks that, in production's opinion, are often not needed – after all the equipment is still running. For maintenance, the term PM means something totally different. For them, it means that work will be done on the equipment, ulti-

The Cultural Infrastructure		
Components	**What to Do**	**How to Do It**
Storytellers	Engage them in the change process	Get the storytellers to tell stories of the new way vs. that of the old.
Keepers of the Faith	Engage them in the change process	Have them provide counseling and mentoring, but focused on the new process.
Whisperers	Engage them in the change process	Work with them so that they filter and pass information supporting the change.
Gossips	Negate the value of their communication	Provide up front and continuous communication to reduce the material to gossip about.
Spies	Negate the value of their communication	Same as the gossips but it will undermine the value of the information they have to pass.
Symbols	Change to reflect the new reliability focus	Alter the symbols to reflect the new focus. Reward those who contribute to the new process.
Language	Change to reflect the new reliability focus	Provide education and involvement so that everyone speaks the same language.

Figure 7-1   Components of the cultural infrastructure

mately prolonging equipment life and avoiding costly failure during operation. Thus, the language and its associated meaning is divisive. This aspect of the cultural infrastructure can inhibit a change for the better.

To break down the language barrier, we need to educate everyone so that the terms have a common meaning. In our new work culture, when people hear the term PM, regardless of their departmental alignment, they will associate it with the elimination of failure, not additional and unwanted work. This adjustment is made through education to promote common understanding. Sharing a common language also helps people in those departments not immediately involved believe that they are part of the program. In our example production being shown how PM directly benefits their work is a start in the right direction. Following this step with ongoing communication about the status and value of the program will further support the effort.

7.10 Why Is the Cultural Infrastructure so Important?

Each of the cultural infrastructure components that we have discussed can be used to promote cultural change or, conversely, to disrupt it. Figure 7-1 identifies each component, providing a brief indication of what and how you need to use them to successfully support your change initiatives.

Changing the cultural infrastructure is not an easy task. Great care and patience must be taken if you are going to make the attempt. However, you must understand that the cultural infrastructure is a hidden force that, if not dealt with, will most assuredly work to undermine whatever changes you are attempting to implement.

The Elements of Change

8.1 The Four Elements of Cultural Change

In Chapters 4 through 7 we discussed the four elements of cultural change. These elements are very important in understanding the organizational culture within a firm as well as what is needed to change it. As I pointed out in Chapter 2, these elements are the foundation upon which we build the soft skills and, above them, the more familiar hard skills such as planning and scheduling, as well as others that address reliability.

The next areas to be addressed in this chapter and Chapters 9 through 16 are the soft skills that I refer to as the eight elements of change. What you shall see in the forthcoming chapters is how the four elements of culture interact with and support the eight elements of change. You will also discover that without a full understanding of both of these sets of elements and how they interact, successful change will be difficult.

8.2 Examples Where the Elements Were Missed

Let us look at some examples where companies made a change but missed a critical element in the change process. These change efforts either failed or, at best, ran into considerable difficulty getting started and maintaining themselves. The bottom line is that, in each case, value for the company was lost. I will provide you with the one line scenario for each and leave the final outcome to your imagination.

Goals are set for a new reliability program without support or buy in by the site leadership

Planning and scheduling are implemented without anyone developing a work process to support their use in the plant.

A revised preventive maintenance program is implemented, but the organizational structure needed to sustain it has not been installed.

A change initiative is implemented, but the organization fails to learn from the mistakes of their past, resulting in one more failed effort.

A computer system (new technology) is installed which the vendor promises will drastically improve how work is accomplished. However, user involvement was neglected.

A company rolls out a change process without the proper level of communication to the organization

A new initiative changes the structure of the organization without anyone taking the necessary time to understand the interrelationships being affected.

The plant changes from an individual performance review / reward process to one that is based on teams, but the teams are not fully functional and neither is the process.

What I have just described are the potential problems associated with failing to take into account the eight elements of change - leadership, work process, structure, group learning, technology, communication, interrelationships and rewards.

Without due consideration of each of these and how they affect the change process, you are inadvertently throwing barriers in your own way. Failure to recognize these barriers sets you up for failure before you even start the change process, increasing the difficulties inherent to this form of work.

The eight elements of change are not independent of one another. Yes, they can and should be addressed as individual topics. But what makes this whole process even more difficult is the strong interrelationships these elements have as well.

Let us examine this concept using as our example the implementation of a planning and scheduling process in a plant site. Assume that you are promoted and assigned to a new plant. One of the primary reasons for your promotion is your successful implementation of a planning and scheduling process in your prior position. Based on your success in the planning arena, you are moved to a new plant site that does not have a planning process, but with the intention of establishing one. However, the personnel in your new plant are not very open to conducting their work based on a planning process. They would rather react to whatever problems are identified by production on a day-to-day basis.

Prior to your arrival, the plant operated in a totally reactive manner.

Although this manner of handling the work was less than efficient, it was widely accepted by those on site as "how we work around here."

Nevertheless, senior management recognize that reactive work processes are less than effective; they have directed you to make the change. With this directive in hand, the first thing you do is to form a planning group, then proceed to force the planning process on the reactive organization. You assume that the plant will recognize the value of your efforts and things will immediately get better.

This unfortunately will not be the case. There are several reasons for this all directly linked to the eight elements of change. One primary reason for the failure is that you did not get site leadership to buy into your initiative. Remember that this new site is reactive; the leaders have been rewarded for this manner of operation.. Next, you did not develop a work process to support the new planning effort. To make matters even worse, you never tried to get process input from the plant personnel most affected by the change. Furthermore, except for your implementation of a planning group, you did nothing to alter the rest of the structure to support the planning initiative. The work teams are still structured so that the foremen respond to the operator's daily direction - certainly not a structure that focuses on planning and scheduling of the work.

On a more personal level, a revised process focused on planning would certainly cause problems with the existing interrelationships. Again this factor was not considered and caused major disruptions in the plant. Further complicating matters, the planning effort was not well communicated, leaving everyone totally confused about what to do and how to do it. The final straw was that you told your new planning team that their yearly performance review and salary increase would be based on how well they implemented the new planning effort.

This scenario reads like your worst nightmare and indeed it is. However, as bad as this seems, it is not far from reality. Every day, plants experience this level of difficulty in implementing change because they fail to address the eight elements of change both individually and collectively.

8.3 The Eight Elements of Change

An essential set of eight elements must be addressed if you wish to successfully address change within your company. Failure to address these elements can lead to problems and ultimately failure of what you are trying to achieve. These elements are:

Leadership

Work Process

Structure

Group Learning

Technology

Communication

Interrelationships

Rewards

Although the elements are each unique, as a group they must work interdependently if you want to implement successful change. They make up what we referred to in Chapter 2 as the soft skills of the change process (see Figure 8-1).

At the foundation (the first level) is the cultural link. In Section 8.4, we will discuss how group learning ties in to the four elements of culture; you will then clearly see how these elements and those of culture are closely inter-connected.

At the next level are the enablers. These elements, leadership and technology, build on the foundation of learning. They enable the group to change successfully. I have seen many thoughtful change initiatives fail because the leadership either didn't understand them, didn't want change to happen, or were comfortable with the status quo. I have also seen changes take place because the leadership strongly believed in and

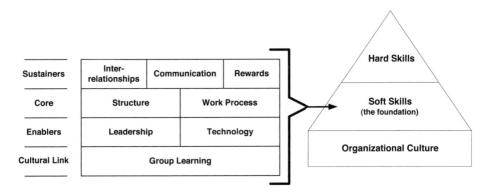

Figure 8-1 The four elements of culture as the foundation of change

led the change.

Technology is also an enabler. With cutbacks and smaller staffs, both information and ways of completing jobs in less time are important. These enablers—good leadership and sound technology-based solutions—support other components of the change process.

The next level of the structure, made up of structure and work process, is the core. Structure defines how you are organized to accomplish the work whereas work process defines how the work actually gets accomplished. When you develop your change initiative, these two elements are essentially what you are going to change.

Once you have worked through the core elements, the remaining elements are the sustainers. The long-term success rate of any change is dependent on communication, interrelationships, and rewards. Communication focuses on how information is transmitted from those who have it to those who need it. Interrelationships determine how well people within the company get along and work cooperatively toward a common goal. Good relations can not by themselves overcome other problems and deficiencies. However, bad relations can undermine the best of plans. The reward system is the last of the sustainers. In many cases, the reward systems of the past will not sustain change. We need to create new reward systems based on a work team approach if we wish to sustain the change we have created.

8.4 Group Learning - The Connection

It is important that we understand the connection between the eight elements of change and the four elements of culture. To make a change successfully you must address each of the eight elements with each of the four elements prominently in your mind. Failure to do so will lead you to frustration, difficulty beyond what is necessary, and in many cases failure in the very initiatives you are trying to implement.

Group Learning is the key element that connects these two sets of elements. If we have in place a learning organization, then we can take the time to learn about our organization's culture at the level of values, rites and rituals, role models, and cultural infrastructure. We can then properly apply what we have learned to the eight elements of change, thereby setting ourselves up for success.

Let us take a look at the two types of group learning. These will be discussed in greater detail in Chapter 13. Figure 8-2 shows Type 1, what is referred to as single loop learning. In this case we set our goals, con-

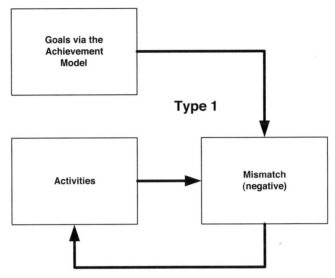

Figure 8-2 Type 1 learning

duct our daily activities, and compare the outcomes with the goals we have set. Where discrepancies exist, we have a feedback loop into the activities. We can then adjust the activities to bring ourselves into alignment with the goals.

Type 1 learning has an inherent problem. It assumes that the goals you have set for organization are correct. But suppose they are not. Suppose the organization's goals were to respond to the needs of production in a timely manner to restore plant operations to normal. This is reactive maintenance. Unfortunately, Type 1 will never let us improve this work process. The reason is that our gaps will only show how we are doing against the goal of better reactive maintenance. Therefore, when we address the gaps via the feedback loop, we only serve to improve our reactive processes. For example:

> If our response time isn't fast enough, we will figure out how to speed it up.

> If we don't have materials on hand, we will increase our stores levels.

> If we have to relocate the workforce from another plant area, we will realign ourselves so that mechanics will always be in each area.

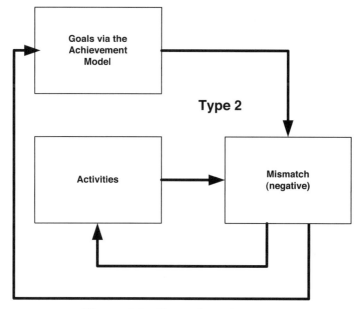

Figure 8-3   Type 2 learning

Type 1 learning doesn't always help us improve; it only helps us get better at what we are already doing. If what we are doing is the wrong thing, then we get better at doing the wrong thing, and fail to improve.

Type 2 learning, called double loop learning, addresses the problems of Type 1. In this model we recognize that our goals may often be incorrect. Therefore, we provide ourselves with a second feedback loop so that goals can be changed. This model is shown in Figure 8-3.

By being able to alter our goals once we recognize that they are not delivering what we require of the organization, we have provided ourselves with a way to improve. But recognizing that we want the organization to improve is only scratching the surface of a far more complex and challenging problem – how to make it happen. In order to be successful at making this change, the organization's culture must be addressed.

8.5 An Example – Reviewing the Four Elements of Culture

Let us look again at the earlier example in which you have been promoted and sent to another plant to implement a planning and scheduling process. In this example, I described how you failed to take into consideration the effects that the eight elements of change would have on this initiative. As a result, the initiative you were trying to implement most likely failed.

Suppose you had carefully considered and addressed each of the elements before proceeding with the rollout of the planning initiative. What do you think would have happened? My guess is that you would have seen some initial success. However, in the end the good ideas you brought to your new assignment would have disappeared. The plant would have returned to its former manner of operation. This would be the case especially if you were promoted to a new assignment or given a major project that lessened your focus on the initiative.

An experience of this sort is very frustrating to those involved. They feel that for some unknown reason a valuable change for the better has been lost. What is even more frustrating is that, as the champion of the effort, you can clearly see the benefits yet have a very difficult time understanding why those around you can not see the same thing. The reason behind this frustration is the failure to change the organization's culture before the attempt to implement the change.

Let us go back to our example once more, but this time begin the process by addressing the four elements of culture as a first step to making the change. Going about a process change in this manner will take more time. However, the likelihood of success is well worth the effort and perseverance needed.

After you arrive at your new assignment and get settled, the first thing you do is not announce the new planning process. Instead, you begin an inquiry to determine the culture in the plant. To do this, you need to carefully consider all that you have learned in Chapters 2 through 7. You need to ask and have answered questions specific to the four elements of culture. Such questions include:

> What are the values that the organization holds relative to main tenance, reliability, and work planning?
>
> What is the role of each of these activities within the plant?
>
> If the plant is reactive, as you have been led to believe, is the organization ready to learn using the Type 2 model and change its goals?
>
> Who are the role models emulated by those in the organization?

What are the rituals that drive the work processes within the plant?

What rites exist that reinforce these rituals?

What does the cultural infrastructure look like?

Who are the key players who influence "how things are done around here" on a daily basis?

These questions are difficult to answer. They will take a lot of research and time on your part. But without this knowledge, you should understand that a planning initiative, or any other initiative, has a far better chance of failure than it does for success.

Let us assume that the plant's leadership has already engaged in Type 2 learning and has recognized that they need to change to a reliability-focused work culture, of which planning is a critical part. That would explain why you were brought into the plant with your skills in this area. In addition, your research has helped you to identify the organization's role models. You also clearly understand the rituals and rites that support the current process. From observation, you also believe you have a good understanding of the cultural infrastructure.

Armed with this information you are now ready to proceed and begin the process that will bring maintenance work planning to your plant. You are now ready to address the eight elements of change with the information you have about the four elements of culture. Figure 8-4 recreates Figure 2-5 and depicts the interrelationship of these elements.

Identifying the issues that need to be addressed within all of these relationships is not as hard as you may think. However, I would suggest that the exercise not be handled alone. Having a team from the site working together is preferable to doing this exercise in a vacuum. Not only will you get the benefit from the team's experience, but you will also be helping them recognize their own barriers that have to be overcome if the effort is to succeed.

Ask the following questions for each of the four elements of culture as each one relates to the eight elements of change. The questions should be asked for each block of the matrix in which an "M" indicates a strong connection between the two. It should also be asked of the blocks in the matrix represented by "m." Even though these relationships are not as strong, understanding the connection is important.

	Values	Role Models	Rites & Rituals	Cultural Infrastructure
Leadership	M	M	M	m
Work Process	M	M	M	m
Structure	M	m	M	m
Group Learning	M	M	m	m
Technology	M	M	M	m
Communication	M	M	M	M
Interrelationships	M	M	m	M
Rewards	M	M	m	m

M = Major interaction   **m** = minor interaction

Figure 8-4   Comparison – eight elements of change vs. four elements of culture

What are the issues within the

(select one of the four elements of culture)

that impacts

(select one of the eight elements of change)

What needs to be changed to bring them into alignment
with what ibeing implemented?

Let us see how this works by taking a few examples from our discussion of implementing a planning process in your plant. For this analysis, we will look at the following four combinations;

- Values as they impact *Communication*
- Rituals and Rites as they impact *Technology*
- Role Models as they impact *Interrelationships*
- Cultural Infrastructure as it impacts *Structure*

Examining these four combinations should give you an idea of what is required to analyze and understand the changes that are required for a successful outcome.

Values and Communication

What are the issues within the plant's value system that impacts communication? What needs to be changed to bring them into alignment with what is being implemented?

Currently the value system is supportive of reactive maintenance. As a result, what gets communicated is not "what are you going to do to repair the equipment so that it doesn't fail again," but rather "when will you be finished" or "don't waste time analyzing the failure – just get it back in service." If the plant is truly going to support reliability, the communication from management will need to be changed to support the new value system. Otherwise, the site personnel never will alter their maintenance strategy.

Rituals, Rites, and Technology

What are the issues within the plant's rituals and rites that impact technology? What needs to be changed to bring them into alignment with what is being implemented?

In a reactive work environment, technology (e.g., the supporting computer system) is not often fully utilized in order to get the work accomplished. There is usually a lot of emergency work and often no work orders to back up the request. The materials required to do the work are often not listed or wrong. There is little or no prior work history. As a result the system is unable to support the work process. If the site is planning to change to a reliability focus, the system will need to play a far more important role. To do this, the site would most likely need additional training and a review of the available functionality so that it can be fully utilized.

Role Models and Interrelationships

What are the issues within the plant's role models that impacts interrelationships? What needs to be changed to bring them into alignment with what is being implemented?

In a reactive work culture, the maintenance foremen and the work crews typically do whatever is required by production with little or no planning. These interrelationships work fine in a reactive environment, but are detrimental to sound reliability work processes. Moving forward with a reliability initiative, including planning, will require that these relationships be broken.

Cultural Infrastructure and Structure

What are the issues within the plant's cultural infrastructure that impacts structure? What needs to be changed to bring them into alignment with what is being implemented?

Because you will be implementing a new planning structure within the current organization, you need to address how the cultural infrastructure can support it and, at the same time, avoid areas where it can subvert the structure. As part of the change process, you need to utilize Keepers of the Faith (mentors and protectors of the new culture), Whisperers, and Gossips so that information supporting the new way of working will be passed through the informal organizational structure. You also will need Spies so that you know what is really going on and can, in turn, take proactive corrective action.

This analysis is time consuming and challenging. It requires a new way of thinking about change. However, to successfully implement change, the organization's culture must be understood and taken into account during the pre-implementation phase.

8.6 Addressing the Eight Elements of Change

In order to provide an in-depth knowledge of how to undertake this analysis, each of the eight elements of change will be addressed separately in Chapters 9 through 16. Each chapter is dedicated to one of the elements and will describe the element in detail, addressing each elemental relationship as depicted in Figure 8-4. Armed with this information, you will be able to understand and conduct your own analysis of your situation.

If you have read my book *Successfully Managing Change in Organizations: A Users Guide*, you will remember the Web of Change and the associated survey. This survey was designed to give you an idea of where the strengths and weaknesses were within your own change process. For more detail on this subject, see Chapters 10 through 13 of that text.

In this text I have taken the next step. Chapter 17 introduces a new web model called the Web of Cultural Change. This model is similar to that of my previous book. It uses the eight spokes of a radar diagram to show the change status relative to the eight elements of change and, within each, to the four elements of culture. Chapter 18 further supports your effort in that it introduces the concepts of assessment of the

Web of Change and change root cause failure analysis. These processes ask you questions related to the eight elements of change and the four elements of culture. It also provides you with a way to assess your ability to sustain your initiative over time.

Leadership

Part 1: The Basics

9.1 Introduction to Leadership

The first of the eight elements to be discussed is Leadership because, in my opinion, it is the hinge pin of the entire effort of changing an organization's culture. It is possible to overcome problems and even outright deficiencies in the other areas, but not possible in the area of leadership. Organizations can develop a vision, create goals and initiatives, and begin processes and projects, but without leaders, change efforts will fail, even those with the potential to succeed. It is the organization's leaders who can see the end state and, through their special set of skills, help the organization attain its goals.

In Part 1 of this chapter, I will set in place a very general groundwork of leadership. Then in Part 2, I will discuss how leadership relates to organizational management and the process of cultural change. From the topical headings of Part 1's sections, you can see that my first goal is to address the more practical side of the subject, showing how proper leadership can make your effort successful.

9.2 Why Leadership Is the Most Important Element of Change

You should understand two definitions as we begin this discussion: leadership and management. As you will see, these terms are not mutually exclusive. In fact, in the conventional definition, leadership is portrayed as a component of management. Although the distinction between leadership and management is often blurred, it is entirely possible that the two functions may not be embodied in the same individual in your company. In short, the differences are significant.

Management has been characterized as "doing things right". In this simplified definition, things includes the four traditional functions of management: planning, organizing, controlling, and leading. Notice that leadership is one of the components of management. If you think about how all four components fit into your daily work, you will see that a conventional manager is expected to:

- plan what is to take place
- organize to accomplish the plan
- control what is taking place
- lead the group toward the planned goal

In many ways, this definition, although simple, applies for the day-to-day work effort. Conventional theory believes that, in addition to performing the other three components, you are also expected as a manager to lead your organization towards the accomplishment of its goals.

The second term, leadership, is conventionally defined as "influencing others to accomplish the goals of the organization." Even though leadership is a component of management, it is very different from the other three components. The key difference is the word influence, which implies getting people to do things, often without having or using direct authority over their actions.

The other parts of the management definition are focused more towards specific tasks that accomplish the goals. Influence, which is not always task-specific, is very different. This difference has major implications when trying to understand how the element of leadership fits into the process of cultural change. Although the definitions are linked, management's "doing things right" is not the same as leadership's "doing the right things."

Suppose that your organization was directed to install a new computer system in your office. As the manager, you put a team together (organizing) to accomplish the task. The group meets, gets a full understanding of the scope of the project, and builds a work plan (planning). They then meet with you, review and upgrade the plan (controlling), and begin the effort. Because this project is so difficult, you hold weekly status meetings to make certain that the team is on track and has resolved any work-related problems (more controlling). As the manager, you assigned the team and communicated the expected results. These steps constituted a form of leadership. Your influence is embodied both in the consequences of not completing the task and in the rewards associated with doing it well.

This example has a clearly defined set of tasks. Something needs to be done and the manager takes the correct steps to make it happen by the required completion date. The steps in this type of project are very easy to define. In reality, there are very few alternative methods to accomplishing the end result. Furthermore, if the group is falling behind, it will be caught at the weekly meeting and corrective action taken. In this example, the leadership role is overshadowed by the other roles that the manager uses to address the process. In fact, once the course has been set, the person appointed as the group leader will be exerting the majority of the leadership function in the effort.

Now let us look at a different example, one that addresses the process of cultural change. Suppose that your organization is not operating profitably. In a closed door meeting, your manager clearly spells out that if your organization doesn't improve, it will be closed down. You have one year to turn the business around. Because we've made this is a simple example, let's assume that if you can fix the organization internally (change the work process), then conditions in the marketplace will allow for profitability.

In this example, the ultimate goal has been set: stay in business. A leader, your manager, has definitely influenced you. However, there is not one clear path, nor is there one correct way to accomplish the end result. In fact, the end result may not be clear either. If you structure your effort simply to become a profitable firm in the next year (short-term focus) and don't plan for the future (long-term focus), you may find yourself in several years back where you are now.

The question is: Within the context of the conventional definition of management, what do you do to achieve success for yourself and your company? The planning, organizing, and controlling components of management still apply, but in this case they are vastly overshadowed by the component of leadership. In fact, you don't have to be the one who plans, organizes, or controls. However, if you want to have a job at the end of the year, you had better be the leader.

Based on this example, I see a need to define leadership specifically to the process of cultural change or change in general. In this scenario, management has two very distinct parts: *managing,* which includes planning, organizing and controlling; and, separately, *leadership.*

My definition of a change-management leader is a person who can clearly see and communicate a vision of the future, influence others to embrace the changes necessary to accomplish that vision, maintain organizational focus over an extended time period, and support group learning as the process evolves. It boils down to "what we conceive and believe, we can achieve," The leader is the person who can pull it all together.

Between our definition of management that is focused on the task components, and our definition of change management leadership, we can see that in the arena of change, the manager doesn't always lead and the leader doesn't always manage. This conception sets the stage for future discussion.

9.3 Some of the Basics

The next step in understanding leadership is to look at what I consider the most pertinent theory on the subject. As your change process proceeds, there will be times when leadership is an issue and understanding some of these basics will help.

Working for many years in an effort to change organizational culture, I have worked for some great leaders, some average leaders, and some non-leaders. In each environment, I had some difficulty because when what seemed to be a natural and correct course of action failed, I didn't understand the influence that the leaders had in causing these outcomes. Knowing some leadership theory will help you ask the right questions to analyze a situation. This information will help you identify and get past the leadership issues that can bring your process to its knees.

Transactional and Transitional Leaders

Essentially, there are two types of leaders. Transactional leaders motivate groups to perform at expected levels and transitional leaders motivate groups to perform beyond expected levels. In the case of cultural change management, we need transitional leaders. Change management requires its own definition of leadership. In the area of change, the leadership function overshadows the planning, organizing, and controlling components of management. The conventional definition of management doesn't fit our needs, but more clearly fits the transactional leaders.

By reviewing the two definitions, you should be able to see how the leadership in your company fills the transitional and transactional roles in the process of change. Transactional leaders may support the change process, but their leadership ability is limited to getting the day-to-day

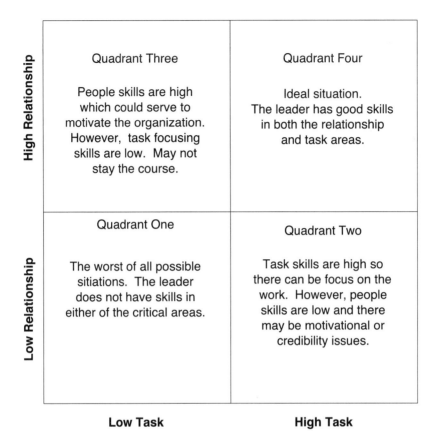

Figure 9-1 Leadership task and relationship quad diagram

things done. What we require are transitional leaders who have the ability not only to see a different end state, but also to influence others to move towards it. This distinction becomes very important when we discuss management / leadership organizational mismatches.

Tasks and Relationships

Another equally important concept looks at how a leader relates to tasks and to people. The task component addresses the extent to which a leader spells out what to do. The relationship aspect focuses on how well a leader engages others in the process. Figure 9-1 shows the relationship between tasks and relationships.

What we really seek is a leader who does both well. The problem is that this combination of leadership skills doesn't always exist. Some fortunate firms have leaders with both, and if yours is one of them, then you are extremely lucky. However, the combination doesn't exist in most leaders.

You can see from Figure 9-1 that the ideal leader has both a high task and a high people orientation (quadrant 4). The worst leader (quadrant 1) fails to possess either set of skills. Leaders in that position are unlikely to last very long. Unfortunately, if they do, the change process won't. The other two combinations are equally undesirable because the set of needed skills is out of balance. A leader with more task and less people skills may fail to provide the proper motivation (quadrant 2), and a leader with more people and little task skills may be viewed as superficial, always focusing on relationships, but not moving the work process forward (quadrant 3).

Although this analysis is not in-depth, if you look at your leaders, you will probably be able to fit them into one of the quadrants. Their position in this grid may help to explain to you why they do what they do. More important, their position may help you not only to categorize their deficiencies, but also to provide them with support needed for their weaker areas. This support may very well be what will sustain or possibly even save your change process from failure. I can tell you from personal experience that being able to support your leader's skill deficiencies is a valuable addition that you can bring to your organization. Suppose that your group leader possesses strong transactional skills. This will result in a great deal of day-to-day work being accomplished. However, the transitional component needed for continuous improvement may be lacking. If you can provide this to the group, then both transactional and transitional work initiatives can be delivered.

Power

We have now talked about the dual orientations of task and people, and the importance of having both skills. Next, we look at the relationship between the leader and the members of the organization or group who are engaged in the change process.

Especially in the area of change management, leaders need to make major changes if they expect to achieve the far-reaching goals usually associated with this type of effort. The problem is that not all leaders have the support of the groups that they are trying to lead. There are

Type of Power	Typical Organization Reaction
Legitimate Power - Provided to the person by the organization	Compliance at least on an overt level. How the person acts and uses the power bestowed will determine if the compliance turns to commitment or resistance.
Reward Power - Based on the ability to provide rewards	Compliance in order to acquire the reward, assuming that the reward is something desired.
Coercive Power - Ability to punish for inaction	Resistance is usually the outcome of application of this form of power. People want to be treated with dignity and respect.
Expert Power -Based on possession of information	Commitment so long as this form of power is not abused. For change to succeed, information needs to be shared.
Information Power - Access to and control over information	Compliance which can turn into resistance if the needed information is withheld.
Referent Power - Leader is admired, identified with	Commitment because these individuals are usually natural leaders.

Figure 9-2 Types of power

many reasons for this, but in my opinion the root is in the way they got their power and how they used it in pursuing the task.

Power is "the capacity to affect the behavior of others in a desired direction." There are different types of power, some good and some not. Each type, and the way each is used, affects people's behavior differently. Thus, power and its use directly impact the relationship between leaders and their groups. The change process is more likely to be successful when the relationship is positive than when the relationship is poor.

Six types of power apply to leadership. Each of these types, summarized in Figure 9-2, has associated with it one of three typical reactions that could be expected from the group being led. These three reactions

— commitment, compliance, and resistance — represent different types of energy (positive, neutral, and negative respectively) that the organization will expend on making the change process a success.

You can see from the different group reactions that some forms of leadership are beneficial to the change process whereas others have minimal or even negative impact. The degree of commitment that a group has will affect how they handle difficult times in the change process. A leader with a group's commitment will be able to lead them into areas where another group would never venture.

The next type of reaction is compliance. Groups within an organization that are functioning in this mode are not really behind the effort. They may be going through the motions, but belief in the process is simply not there. I believe that it is possible to lead a compliant group to function efficiently and move the process forward, but that leadership takes considerable effort and has a high degree of risk. Such a group may move forward, but only through the will of the leader. The risk here is that when the leader leaves, the process crumbles.

Resistance, the third reaction, is altogether different. In *Successfully Managing Change in Organizations: A Users Guide,* Chapter 9 discussed this subject at length. A change process is doomed if the leader is working from a coercive frame of reference and the organization is resisting. The only quick way to turn this around is to replace the leader with someone who is not functioning in this fashion. For many reasons, this is not always possible. If that change were to happen, however, the new leader would need to address the issue head-on and resolve any "old baggage" from past regimes. In these situations, it is often necessary to let some time pass while the new leader gets acclimated and the former situation cools down before the new leader can again move forward.

9.4 Why Is Leadership Needed?

If you are one of the people in your organization who handles projects or work efforts of varying degrees of complexity and have been reasonably successful over your career, you probably ask yourself, "What leadership do I need? Just give me the project and I know how to get it done." Do you really need influence to achieve the group goals? These are important questions that are often asked by mid-level managers. You need to understand the answers in order to appreciate fully how change works, and why change really needs the leaders that we have discussed thus far.

If you fit the above description, then you already are, in fact, a leader. Depending on the type of effort in which you are engaged, you may be a change leader as well. More likely, you are leading a subset of the overall effort. The question about the need for leaders is aimed at a much higher level within the organization, to the need for an overall leader of the change effort.

There is a need of an overall change leader for several distinct reasons. That leader has the responsibility to unite diverse groups around a single vision for the firm. Although it is distinctly, even highly, probable that all efforts are aimed at a single overall goal, it is not realistic to assume that many individuals or groups acting independently will be able to work together to be successful. Diverse efforts led by many highly skilled individuals need a single contact point, someone who sees not only the overall vision, but also how the individual components should fit together to support the whole.

An overall change leader also has the responsibility to initiate the process of converting vision into company-wide reality. Creating a vision is not an easy task. Taking that vision and creating a framework within which each of the individual change leaders can align their efforts is an even more difficult task.

Another role of the overall change leader is to keep people focused and on course. Change is not an overnight event, but a long, winding road. With so many people making the journey, it is easy to lose focus and drift away from the long-term goal. The overall change leader has the responsibility to prevent this from happening. Because organizations, groups, and individuals lose their focus from time to time, it is critical to help them get it back. Loss of focus is like a virus: once it starts to spread, it spreads quickly, and has the potential to kill the patient.

The last role is that of cheerleader. The overall change leader can not do everything; a smart leader knows to resist the urge to try. The urge to do too much can be a serious problem for the organization if is acted out because it misrepresents the nature of the work. However, a smart leader can stay in the background providing encouragement to everyone along the way. As I have noted, change takes time; people are seldom patient enough to allow the proper amount of time for the change to be successful. A good change leader keeps momentum and maintains the energy, even when it seems to be disappearing.

We have talked about upper level management leadership. Equally important, there is a need for leaders at all levels of the organization. It is insufficient to have a leader at the top and followers in all other levels of the structure. Just as the overall leader has responsibility to the process of change, so does a leader of a department or a sub-group within the company.

The problem is what if the managers are not the leaders? What if there is no alignment between the leaders in the organization? A discussion of these subjects follow.

9.5 Managers Are Not Always Leaders and Leaders Are Not Always the Managers

There is a myth that the manager is always the leader. This especially applies to change efforts because, by their very nature, they are more strategic, and strategy is what a manager / leader is supposed to do. If you have a situation where the manager is in fact the leader, then you are very lucky. However, this is not always the case. There are many combinations of leadership and management that do not fit this ideal model; some even have the potential to undermine the change process.

Before we begin this discussion, we need to clarify who and what responsibilities we mean when we refer to the manager. In all organizations, there are many levels of managers and equally many levels of leaders.

The management aspect can be understood simply by envisioning any organization. Embedded within the organizational structure are various levels of managers. Some organizations actually have too many levels; in others, the levels are few and the organization is referred to as being flat. Nevertheless, managers exist at all of the levels.

Suppose we have a four-tier organization. At the top sits the department manager followed in level by their superintendents, their foremen, and, at the bottom level, the workforce. Our conventional definition of manager applies to any of the three top levels. Managers at each level are responsible for planning, organizing, controlling, and leading (the conventional management definition) within the scope of their own job.

At each organizational level, a manager's job is usually well executed if they possess these defined management skills. This structure works well for day-to-day work efforts in which the first three elements of managing overshadow that of leadership. Recall from Section 9.3 that such managers were referred to as transactional leaders.

When we look at change management, we are dealing with a process different from the day-to-day. This type of effort calls for the transitional leader, armed with a different set of skills than the transactional leader. In this case, leadership overshadows the other elements of management.

Most managers have the leadership qualities described for transactional leadership. They probably got to the positions they hold because they were able to get things done. But in the case of change management, the qualities and skills of a transitional leader are different and more difficult. Unfortunately, the organization usually expects their current mangers to handle that role as well.

What if the manager can be successful in the transactional environment, but does not have the additional and different skills it takes to lead the organization through a difficult change process? What will

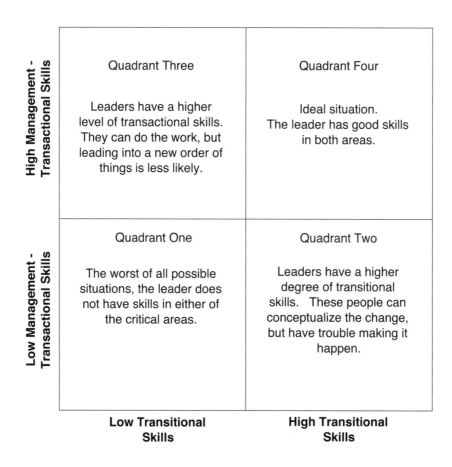

Figure 9-3 Transactional vs. transitional skills in leaders

happen if the manager can not be the transitional leader that is required?

There are four possibilities, as shown in Figure 9-3.

The lower left is reserved for people with neither set of skills. They should not be managers. Although people in this quadrant are managers in many companies, they should be reassigned to positions where they can be of most value. They do not have the necessary skills to help us succeed in the change process.

However, removing or reassigning them is sometimes easier said than done. Those of you reading this may or may not even have the level of authority to make a change of this sort. One thing is almost certain: if these individuals are not replaced, they will still be in their jobs long after the change process has failed.

I once worked for someone who, in my opinion, lacked the skills to lead our organization to a new way of doing our business. He was an immovable force because his superior shared the same history and same work philosophy. We tried and did make some minor changes, but it wasn't until he left the company that more significant change was possible. Change was not an overnight event because we had to wait for him to retire! The down side was that our competition was moving forward even as we waited for him to leave. The performance gap between our company and the competition was much harder to close when we finally could begin.

The upper left quadrant depicts people who are good transactional leaders, but do not have the skills to lead the company through a change process. These managers can still be successful and provide real value to the change process; once the strategic direction has been set, they are the ones in the company who can "get it done". They require two things to be successful in the change management role. First, they need to realize that they are NOT the change management leaders. Second, they need to empower those who are. If they fail the first requirement, the likelihood is that the change will never happen. Transactional leaders cannot make change happen. If they fail in the second requirement, the real change leaders (who may actually work for them) may not be allowed to emerge, ultimately hurting the process.

The lower right quadrant is the area for those who have good transitional skills, but may not be as strong in the transactional area. These are the people who can see the future as depicted by the vision. They are committed to make the change process a success and may in fact have some very good ideas about how to make it work. The problem is that without the transactional skills, they can not successfully manage the change process. The conventional wisdom that managers are the leaders we require is not correct.

The people you need to make your change process a success may, and probably do, exist at many levels in your company. Some may be lower-tier managers; others may not be managers at all. The trick is to find these people, then empower them to help you through the change. The transactional leaders that these individuals work for must let them be the transitional leaders that they are. If one of your managers stands in the way of this, you will lose valuable assets to your change effort.

People who are insecure in management roles have the potential to get in the way of change. They believe that a subordinate with a skill that they do not have reflects a failure on their part. Your job is to prevent such beliefs from stopping the change process. Task teams, special assignments, and pulling people out of their regular jobs to complete a specific task are among the ways you can free up the skills required without threatening those in the management jobs. These strategies will work in the development phase. But when you roll out the changes to the entire organization, the managers need to buy in to the effort. Having them out of the loop may ultimately create more problems. This hurdle needs to be addressed, if not now, then later.

The upper right quadrant is the most desirable, but least likely. In this mode, your current managers have both sets of skills. If you have managers in this quadrant, you are very lucky. My guess is that most of your managers are in the other quadrants. Recognize what you need to accomplish. Then figure out how to mobilize all of your resources to achieve the change management goals.

9.6 Expectations of Our Leaders

Now that we have looked at leadership functions, discussed why leaders are needed, and noted that not all managers are the transitional leaders we need, we can look at what we expect from our leaders. (Notice that I am not equating leaders with managers although they may be one and the same). As your change process moves forward, you will need to find people to act in the change management roles. A list of what you should expect from leaders follows.

What we want from change management is that the leaders are Inspired, with a strong belief that change is required for success

Visionary

Confident that they have the skills to help the organization achieve its goals

Committed to stay the course over an extended amount of time

Empowered to act and willing to empower others to do the same

Not threatened by others who have skills that they do not have

Willing to hold others accountable for deviation from the overall goal

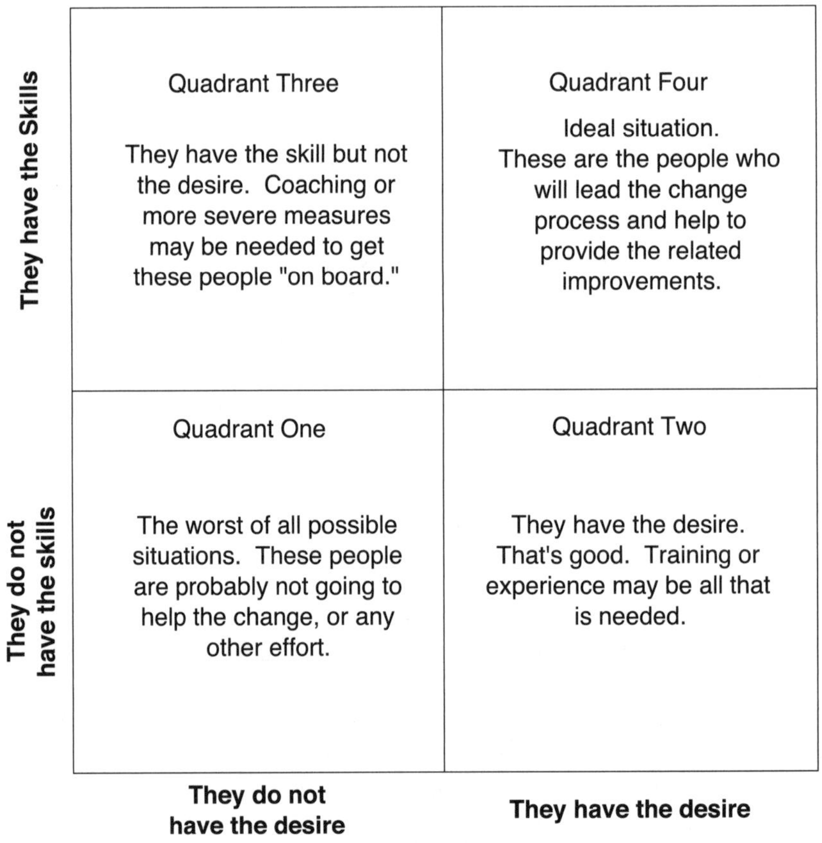

	They do not have the desire	**They have the desire**
They have the Skills	Quadrant Three They have the skill but not the desire. Coaching or more severe measures may be needed to get these people "on board."	Quadrant Four Ideal situation. These are the people who will lead the change process and help to provide the related improvements.
They do not have the skills	Quadrant One The worst of all possible situations. These people are probably not going to help the change, or any other effort.	Quadrant Two They have the desire. That's good. Training or experience may be all that is needed.

Figure 9-4 Skills vs. desire to change

Using this list and adding other characteristics that may be specific to your company should help in being able to articulate the requirements for those who are going to be leading the change effort.

9.7 Leadership Mismatches and Overcoming Mismatch Problems

What if there is a mismatch between what you require of a leader and what a person is capable of or willing to deliver as you proceed through the change process? Figure 9-4 addresses this subject.

In the lower left quadrant are people who do not have either the ability or the desire to do this type of work. They should not be managers, but should instead find work where they can contribute; this may not be within your company. In the upper left quadrant are those who have the ability, but not the desire. People in this quadrant need to find work for which they have a desire; again, this may not be in your company. Those in the lower right quadrant have the desire but not the skill. These people may be organizational sleepers and, once trained, may emerge as valuable company assets for the current change effort and beyond. The key here is to identify these people and spend the time, and sometimes money, required for their development. The upper right quadrant includes those who have both the skills and the desire. They are your "keepers" and the ones who will ultimately be the change agents.

9.8: What Happens When The Leaders Don't Measure Up

This section addresses an issue that is often ignored. It is entirely possible, even likely, that some people will never embrace the change concepts that you are trying to introduce. Recall the Machiavelli quote at the end of chapter one. There are those in your company who did very well under the old system, as well as a great many who are not sure they will do well under the new. You may use various communications methods, coaching, leading, and many other tools you have in your management tool kit, but in the end, there will be those who opt out of the new organization.

There are two ways in which this will happen. The first and simplest is that they will leave the company. Although this is not something you want to happen, I guarantee that some will leave. The second and far more problematic way involves those who opt out, but stay around. They are recognized as the ones who are always telling everyone else that things were better under the old conditions and that the new order will never work. At first, you and your organization should try to salvage these employees, if possible. However, you should consider what to do if, after all of your coaching and leading, they still won't join the change effort. As unfortunate as it may be, once you confirm that their resistance is permanent, you need to help them find other work.

Part 2
Leadership and Change Management

9.9 Leadership and the Elements

Now that we have considered leadership basics, we can turn our attention to how leadership fits into the eight elements of change and how it plays a primary role with the four elements of culture. Your understanding of this topic is very important because leadership is probably the most important requirement for successful change. Without this quality inherent in those who manage your organization, any change program is doomed from the start.

Let us look at some change efforts that failed:

A plant decided to implement a program targeting improved quality. Those responsible for its success were only interested in the form: Did the effort appear to meet corporate standards? The content and actual implementation of a real quality improvement effort was not important.

A preventive maintenance program was put into place, but the reactive aspect of the plant maintenance culture continually pulled the mechanics away from their preventive maintenance tasks; they never returned. Those leading the maintenance department never tried to keep the crew focused on the preventive maintenance program.

Another plant experienced a high level of rotating equipment failure. Because of the high number of pumps out of service, the plant was actually at risk of shutting down several of its major production lines. Management preached reliability, but refused to give maintenance money to get everything fixed.

One company had a major initiative to increase effectiveness and efficiency of the overall work effort. People needed training to be able to make the changes needed to accomplish this goal. However, training never took place because the management said there was no money.

These types of examples could go on and on. If you added your own example, the list would be even longer. A common thread runs through all of these examples: the lack of leadership. Without leadership any change effort is doomed from the start. Conversely, good leadership can develop, communicate, and foster changes that, if asked, one would say could never be accomplished. Good leadership can move mountains!

There are also examples of leadership at its best. At one point in my career, I worked in a company that was getting by, but not really achieving its full potential. The leadership had a vision, but that vision was not clearly and forcefully articulated. Nor was the management team really working to make it happen. There were more excuses why we failed than one could ever imagine. Then we changed management at the senior level. The people brought into the company were leaders. They articulated their vision and began systematically to help the team achieve it. This major change was not easy, but over several years we evolved from what I had considered a mediocre company to one that was recognized as a leader in our industry. We all had wanted to achieve this level of performance, but the leadership made it happen.

The balance of this discussion about leadership will look at it from the perspective of change management, in relation to the eight elements of change and the four elements of culture.

9.10 Leadership and Change

Of the eight elements of change, the most important is leadership. This element has far reaching impact on all of the others. This section examines this impact so that, as change efforts are being developed and implemented at your plant, you will have a heightened awareness of the impact and need for sound leadership.

Three principles are important when addressing the topic of leadership in relationship to change initiatives:

Using Power

Leaders have power. They can apply this power to affect change more than anyone else in the organization. This application of power can be positive or negative.

We have all been in jobs where the leader decided that a change was in order. The leaders may have been caught up in the most recent managerial fad or simply believed a change was needed. In many situations, new leaders are brought into the company, recognize that a change is needed, and then proceed to make it happen. In all of these cases, some form of change will take place. This change happens because the leaders have the power to make it happen. They set the new expectations for the business and then go about working with the organization to meet those expectations.

For example, I worked for a manager who decided that our organization needed to change how it approached maintenance work. It was his opinion that we were too reactive in our approach, did a poor job of work planning and scheduling, and executed the work at a level far below the expectations of our customer. By virtue of his position's power, he created a work process redesign team that spent literally hundreds of hours developing and rolling out a new work process. Someone not in a leadership position could not have set these wheels in motion nor mobilized the organization to get this type of work completed.

The leaders also have it within their ability to make things worse. Think about a leader who enters a company where the maintenance function is based on good planning, scheduling, and field execution. However, the manager determines that the best way to serve production's needs is to put in place a rapid response force dedicated to correcting problems as they happen. Further assume that the manager implements this new concept on a very broad scale, thereby eliminating the former process. A move of this sort would set a proactive organization back years. Because the leader has the power, this sort of change is entirely possible.

Incorporating the Eight Elements of Change

With the power to make change, leaders must recognize the need to use the eight elements of change in a collective and balanced fashion.

Change is multi-faceted, as shown by the eight elements of change. Even if you can change one or more elements and have change occur in the organization, you may still face problems if the effort is not bal-

The Results of an Unbalanced Approach to Change

(missing)	Work Process	Structure	Learning	Technology	Communication	Inter-relationships	Rewards	No direction
Leadership	(missing)	Structure	Learning	Technology	Communication	Inter-relationships	Rewards	No method
Leadership	Work Process	(missing)	Learning	Technology	Communication	Inter-relationships	Rewards	No framework
Leadership	Work Process	Structure	(missing)	Technology	Communication	Inter-relationships	Rewards	No knowledge
Leadership	Work Process	Structure	Learning	(missing)	Communication	Inter-relationships	Rewards	No support
Leadership	Work Process	Structure	Learning	Technology	(missing)	Inter-relationships	Rewards	No understanding
Leadership	Work Process	Structure	Learning	Technology	Communication	(missing)	Rewards	No effectiveness
Leadership	Work Process	Structure	Learning	Technology	Communication	Inter-relationships	(missing)	No incentive

Figure 9-5 What happens when you miss an element

anced. For successful change to take place, an organization must use all eight elements of change in a balanced manner. This approach will provide the greatest chance of success.

A balanced application of the eight elements of change can become imbalanced in two ways. The first way occurs when the leader entirely misses one of the elements. The second way uses all of the elements, but emphasizes one or more of them in too strong a manner.

Figure 9-5 shows the results of the first type of imbalance – missing one or more elements. This figure vividly shows what happens when you are trying to make a change, but have left out one or more of the elements from your change strategy. For example:

Missing Leadership

Without leadership during and after the change, there is no reason to begin. Leadership provides the direction and guidance through the process. Leaders are the only ones who have the power to refocus the organization away from the day-to-day and toward a more strategic effort.

Missing Work Process

Without a new work process, the changes cannot be executed. Furthermore, without this process, the organization will quickly revert to what it knows best – the old way of working.

Missing Structure

Structure must be in place to support the process. Think of the problems that would arise if you developed a process focused on planning and scheduling, but never integrated these positions into the organizational structure.

Missing Group Learning

Without group learning, any knowledge gained as the changes are implemented would be lost. Organizations must learn from their successes and their failures. Leadership can severely hamper this learning process in many ways, resulting in the loss of knowledge and continuous improvement.

Missing Technology

Technology is an integral part of how we work in today's world. Yet some people still resist it. In our efforts to improve reliability, we need information and processes that require state-of-the-industry technology.

Out of Balance Element (too much focus)	Potential Result
Leadership	When overapplied, the organization will feel over controlled and allow the leader to make all of the decisions. If the desire is to implement change throughout the organization, the leader must allow those on the leadership team to develop these skills.
Work Process	Too much focus on the process delivers a good process – on paper. However, the process tends to be somewhat impractical and, when implemented, often unwieldy. It is not simply the process that counts, but the results of the process.
Structure	Too much time spent worrying about the organizational chart's boxes and who goes in which box often disables the process. Change is not about who is in which box as much as it is about what people do when they get there.
Group Learning	Too much emphasis here makes the work too theoretical. Most people in maintenance organizations are doers. They want to learn what it is that they are supposed to do and not have to get deeply into theory.
Technology	Too much focus on this element causes organizations to place the "cart before the horse." Although a good computer system is of great value, it must never be allowed to drive the process.
Communication	Communication is an extremely important element. However, groups that overemphasize this element tend to communicate before plans are finalized. They want the organization to know what is going on. However, as the organization develops new ideas and alters the old, over communication can cause confusion.
Interrelationships	Stressing interrelationships above the other elements will cause a strain if the changes to the work also cause the interrelationships to be changed.
Rewards	Too much emphasis on rewards takes people's attention away from what is really important – the change initiative itself.

Figure 9-6   Results of being out of balance

Missing Communication

Lack of communication in a change process will cause it to fail before it begins. Lack of understanding will cause people to create their own answers to their questions, leading to misdirection and failure of the change effort.

Missing Interrelationships

Without interrelationships, the change process will not be effective. The majority of the work handled by reliability and maintenance organizations can never be accomplished by individuals working alone. This work always requires multi-functional teams; the glue for these teams is their interrelationships.

Missing Rewards

Rewards can not be left out of the overall strategy. Without them, others have no incentive to make the change. The initial thought when discussing rewards is often to think of money. Yet money is not the real driver. The rewards that have value are work content, responsibility, group association, and self-esteem. All of these are missing if rewards are not in the change equation.

The second type of imbalance, where one or many of the elements are emphasized too strongly, causes the organization to spend more time on that element. As a result, other elements suffer from inattention. As a result, the change put in place is lopsided; the elements that received only a small amount of attention during the change process will come back to haunt you. Figure 9-6 shows representative examples for each of the elements.

Incorporating the Four Elements of Culture

Leaders can use the eight elements to make change. However, without also addressing the four elements of culture, any changes they have made will most likely be gone shortly after they leave the position.

We have seen how leaders have the power to effect change and how they need to use the eight elements in a collective and balanced fashion. Yet for leaders to truly succeed, they need to do more. They must also make changes to the four elements of culture. Otherwise, long lasting change will never occur. Change that does not include the four elements of culture will exist only as long as the leaders, through their force of will, keep it in place. If they turn their attention elsewhere, retire, get promoted, or leave their position for any reason, the old process will emerge from hiding and be back in place before you notice.

9.11 The Four Elements of Culture

Our leaders play a critical role in cultural change. We can work very hard to change work process, structure, or the other elements of change. However, if we do not alter the culture at the same time, we will ultimately fail. The elements of change are manifestations of the plant culture or how things are really done around here. As they promote their change initiatives, leaders must at the same time change the culture. Therefore, when you hear one of your leaders speak about cultural change, you should be encouraged. Although they may not know exactly what is involved, the fact that they have recognized that this level of change is needed is significant. In this section we will explore how our leaders impact and ultimately have the power to change the culture.

Organizational Values

The values that an organization employs are set by its leadership. If an organization values reactive maintenance, then this approach will be what is communicated. Similarly an organization that desires that the plant run reliably, always being in a position to deliver its products to meet market demands, will establish a far different set of values. Any organization's leadership has a three-phased responsibility: 1) to set the values for the firm, 2) to communicate these values throughout the organization, and 3) to make sure that the communication is understood and put into practice.

If all that the leadership does is set the values, but it never communicate them, the organization will not receive the direction it requires. Checking that the proper level of communication has been made is easily determined by assessing the value structure of the organization, then comparing it to that which is espoused by the leadership. If the values were properly communicated, there should be a perfect alignment between the two. If not, then the communication chain has problems.

Suppose that at its yearly off site meeting, a leadership team determines that the plant needs improved planning and scheduling of its maintenance work in order to improve both its effectiveness and efficiency. This type of change is certainly something that is not solely the responsibility of maintenance. In fact, in order to have a good maintenance planning and scheduling process, production (the main customer) needs to be heavily involved, as do the other support organizations. Therefore, the leadership needs to communicate this value to everyone and have them realize that they are all part of the solution.

Now suppose that six months after the off site meeting, no progress has been made. Maintenance is trying to implement the new way of working, but production and the other support groups have not participated. When asked why not, they express their belief that planning and scheduling are purely the responsibility of maintenance. In this case, the leadership set the value, but failed to communicate it, thereby setting the new planning process up for failure.

If the leadership sets the values and communicates them, but never makes certain that their communication was understood, then they have not fulfilled their entire leadership responsibilities.

A communication can be misunderstood or misinterpreted in many ways. As a result, employees can do the wrong thing while believing that what they are doing is exactly what they were asked. It is insufficient for leaders simply to communicate. They must also insure that the communication was received, properly interpreted, and acted upon. They need to undertake periodic independent audits of the process. If these audits reveal that things are working as expected, then all is well. However, if there is a mismatch, then the evidence indicates a lack of understanding that must be corrected.

Role Models

Leaders must be aware of those within their organization who are role models. The culture emulates these people. If they represent the values of the business and support the goals and initiatives associated with change, then all is well. However, quite often when a company is making a change, the role models represent the old way of doing things. This conflict can be very detrimental to process change because the organization, left to its own devices, will continue to follow the role models and the old ways.

A leader can do many things to alleviate this situation; they will be discussed in subsequent chapters. However, it is important for leaders to understand that a failure will occur unless role models who support the old way of working are taken out of the process and away from their supporters.

Suppose we are going to implement a process that applies strict planning principles to the major plant outages. These are the major events for which continuous process industries such as power generation, petrochemical, and others remove production units from service in order to perform repairs that can not be addressed while the equipment is in operation. However, those who have been in charge of these efforts in the past apply more of a "seat of the pants" approach. This approach has been successful until its total cost in downtime is really examined. An analysis indicates that this approach is very poor. However, the organization never sees these details. All it sees is what appears to be outage success stories. As a result, those currently in charge become role models for these efforts. If the leadership wants to implement a significant change to this process, it needs role models who will support the new approach. Ideally this support will come by those presently in power accepting the new process. Otherwise, managers have to be found to be the new role models for the new way of doing business.

Rites and Rituals

Leaders control rituals by how they implement the changes they make to the organization and to the process. These changes become the new rituals by which the organization executes the day-to-day work. Leaders also control the rites that are associated with these rituals. If praise is given to those who execute planned work, then the ritual (planned work) and its supporting rite (the praise for the job well done) will become embedded in the culture. Failure on the part of the leadership to make these changes will hamper the culture from accepting new ways and moving forward.

Cultural Infrastructure

The cultural infrastructure exists within every organization; it is comprised of different types of individuals with various roles. It is difficult for leaders to control the cultural infrastructure, which will form and take on a life of its own as it works towards maintaining the status quo. Although leaders can not control it, they must understand it and, when possible, use it to foster change. They need to know who are the story tellers, the keepers of the faith, the whisperers, gossips, and spies, then act in a way that will convert the people in these roles from advocates of the old to supporters of the new.

Work Process

10.1 What is Work Process?

This chapter addresses work process, the second element of the eight elements of change. When we think about this element, we invariably think about the flow of work through a system. This flow consists of creating the work, passing it along to others who do prescribed parts of the job, and then acting to complete the effort. This concept of work process can be applied to the complex series of events required to build a car on an assembly line to the much simpler events required to pay an invoice from a vendor. In each case, the process has the same three components – a beginning, a series of steps where things are done, and completion.

This overview of the most common type of work process defines a specific work flow. Two other types of processes also need to be discussed: information flow processes (or how information travels through the work environment) and communication flow processes. Information flow processes will be discussed in this chapter, communication flow processes in Chapter 14, which deals exclusively with the topic of communication.

In the cases of work flow and information flow, both are important in how we do our business now and even more if we are going to create a reliability work culture. The reason is that rites and rituals, which are part of every work culture, are based on these two flows. Consequently, if you chose to modify the culture to one that is reliability focused, you will need to change the rites and rituals, as well as the flows that support them, in order to be successful.

10.2 What Is a Work Flow Process?

This first type of process describes how work flows through a system of people who at each step change the state of the work until it is completed. By change the state I mean that they do something to the work while they have it in their possession that changes it to something different after they are finished. Often these changes of state are referred

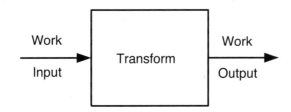

Figure 10-1 Basic transformation

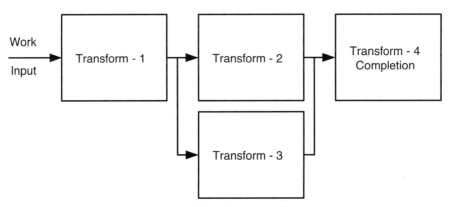

Figure 10-2 Complex transformation

to as a transformations or transforms for short. Figure 10-1 shows a single step in a work process.

In this figure, there are three steps: 1) a work input step, 2) a step when the work is *transformed* into something else by whatever department or group is handling it, and 3) a work output step. For example, take the performance of a planned maintenance task. The input is the work plan, the transformation is the execution of the plan, and the output is the completed work. Whatever enters the block marked transform has something done to it and leaves differently than when it entered.

In the handling of reliability- and maintenance-related tasks, no work flow processes are as simple as the transform diagram described in Figure 10-1. Instead, we have to deal with complex transforms in which work flows along many pathways, some serial and some parallel,

Input	Transform	Who Handles	Output
Equipment in need of repair	Production submits a request for work	Production	Work request
Work request	Accept and convert to a planned work order	Planning	Planned work order
Planned work order	Resources allocated and work scheduled	Scheduling	Scheduled work order
Scheduled work order	Work executed	Work Execution	Completed work for production review
Completed work	Acceptance	Production	Job completed (no further steps)

Figure 10-3 Transformation process of a work order

as shown in Figure 10-2.

Staying with the same example, let's examine the actual number of transforms that a work request could go through on its way to becoming a completed work order. Note that in each case the output of one transform is the input for the next. In order not to complicate the description too much, I will avoid parallel paths, but be aware that they exist.

As you can see, the work request goes through several steps along its way to being a completed block of work. It is acted upon by many different groups as it progresses through the work flow system.

10. 3 Efficient and Effective Work Flows

Work flows can be both efficient (we do things right) and effective (we do the right things). Both of these aspects of the work flow are important if we want the work to be handled in an optimal manner. In a plant with a reliability focus, the work will be well planned and scheduled (efficient) whereas in a plant with a reactive focus, work is based on managers who want everything fixed now. In this latter case, production is continually submitting emergency orders (inefficient), interrupting the planned work. In the plant with a reliability-based focus, we would expect that the repairs themselves would be reliability based. Equipment would be returned to service after failure analysis and upgrades to prevent reoccurrence (effective). Otherwise, maintenance would simply make the quick fix just to get the equipment back on line (ineffective).

Being able to examine our work flows critically with an eye to effectiveness and efficiency is very important if we want to be successful making work process changes that will improve the business. We will address these concepts again in this chapter when we discuss how to change a work flow and the culture that surrounds it.

10.4 What Is Information Flow?

Have you ever worked on a job where you needed specific information, but either it did not exist or you could not get access to it? Undoubtedly you will answer yes; at some point in their careers everyone needs information that they can not get. When faced with this problem, you either waited to get what you needed or, if the work effort allowed it, you made a decision based on your best knowledge of the situation. This example shows an information flow process that is broken. In today's world of electronic information management, such a scenario should be considered unacceptable. However, being asked to make decisions without access to the information needed is often the case.

In work flow, physical things move from person to person and are transformed into their final product. Information flow is different. Information flow, generated as the result of the work process, is stored in various systems in ways that make it available to those who need it for their day-to-day business decisions. Some pieces of information are needed immediately while other pieces may not be needed for years.

An information flow model needs to exist as an important part of the overall process. In this way, the proper information is available when needed for making sound information-driven business decisions. In other words, the information needs to be in the right place at the right time. How it does or does not get there is the problem to be solved. Here is where the information flow and the work flow models overlap because information is needed to directly support the work flow model.

Several years ago, I was involved with a plant outage during which many of the pieces of equipment were to undergo repairs. One vessel that we inspected needed extensive internal repairs. We had been working on these for several days when one of the engineers asked, "Why are you spending time fixing this vessel? I condemned it during the last outage." Needless to say, we then stopped and purchased a new vessel. These steps delayed start-up. They also cost us more money because the new vessel needed to be expedited. These errors resulted from a poor work flow and a poor flow of information. Not having critical information in the planning stages of the job (as a result of these two broken processes) cost us both time and money and also generated a great deal of frustration.

Suppose you have the following process: During the work order close out, the area engineer is required to enter comments about the job and what actions are needed the next time the equipment is taken out of service. Now suppose that, in making these comments, the engineer identified a new type of attachment to be installed, why it is being installed, the order number for the attachment, where it is located in the warehouse, and a detailed procedure for installation. With this information, what do you think should be part of the work plan the next time the equipment is taken out of service? The answer is the installation of the new attachment because the right information is located at the right place and the right time, when the repair order is entered into the system. Had this information not been handled properly, the work to install the new attachment may not have even been identified. Such an oversight can be serious if the equipment is taken out of service only once every three years and the attachment could save $100,000 per year.

10.5 Integrating Work Flow and Information Flow

Before we move on, it is important to understand the linkage between work and information flow. Information flow is the result of work flow, but how do these two types of flows integrate with one another? The answer is in their touch points, the places in the work flow model where the information flow provides the essential information for the types of reliability-based decisions we want to promote.

Take the example created in Figure 10-3. This work flow does not require a computer system. Everything described can be handled with a paper system, perhaps using multi-part forms. The touch points to this type of system are few and far between, mostly consisting of the information that is maintained in the planner's files.

Most systems today, however, are far different. Computer systems provide us with access to the information we need as long as the infor-

Workflow

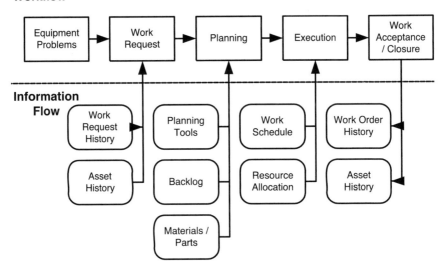

Figure 10-4 Synchronized work flow and information flow models

mation flow model is closely synchronized with that of the work flow model. Then when we need the information, we can quickly get it to support the work flow. An example of a simplified, synchronized maintenance work process is shown in Figure 10-4. The transforms of the work flow are shown as rectangles and those of the information flow as rounded rectangles.

As you can see in this diagram when the work flow and information flow are working together, the information contained in the system is accessible; it supports those who have to make decisions concerning the work. Conversely, it is easy to visualize what happens if the two flows are not working together. Information needed throughout the process is not available to the decision makers, resulting in poor or even incorrect decisions. Therefore, when we look at how to change the work process, we will include both work and information flow in the discussion.

10.6 What Is Communication Flow?

In addition to the work flow and information flow models, there is also a much more subtle process that is often overlooked. This process deals with how we do or do not communicate across the system. We can have the best work flow and information flow models, but if we do not communicate with each other, these models will fail to deliver the required value to the business that we expect. Without communication, we end up in what I refer to as a cubicle world. The departments (or even individuals) each sit in their own cubicle, do work, and then toss the work over the partition to someone else to handle the next phase. This lack of communication can cause serious problems. It is such a significant subject that it is handled as a separate element of the eight elements of change in Chapter 14.

10.7 Repair- and Reliability-Based Work Process Characteristics

It is important to recognize the characteristics of both repair- and reliability-based processes because they both exist in every company. We want to get rid of those process that drive us to a repair-based mind set. To do this, we need to recognize these processes when we see them. They include:

No process – Everything is done as requested and everything is requested to be done now.

No long-term focus – Fix it now and worry about the next failure when and if it happens.

No planning and scheduling of maintenance – This system leads to sub-optimal performance.

Emotional changes to the work effort – This approach results in unplanned interruptions and attempts to execute the work without proper preparation.

These characteristics all have one thing in common. There is a great deal of focus on making the "quick fix" and returning the equipment to service. Working in this manner assures inefficiency and in effectiveness and guarantees that the equipment will fail again – probably sooner than desired.

A reliability-based work process is vastly different. It includes:

A process to analyze the reason for failure and address the root-cause, thereby eliminating this failure mechanism in the future.

A detailed work plan to assure that all repairs and upgrades are handled as part of the work process.

A job schedule so that both production and maintenance can be ready to do the work. Scheduled work allows overall performance to be optimized.

A communication process so that all of those involved know what is to be done, how it will be done, and when it needs to be done.

A post-audit process so that the repairs can be evaluated to assure that everything that was to be done was done and done correctly.

A process whereby the information about the job is saved and accessible for others who may need this information at a future date.

10.8 Work Flow and Information Flow Processes and Culture

Of the four elements of culture, two (values and role models) are required if you want to change the work process and a third (rituals and rites) is integrally linked with it. It will be impossible to implement a reliability-based process if the values of the organization are not focused in this direction. If the values are directed toward a reliability-based process, then performing work using a reactive process will be extremely uncomfortable for the organization, quickly resulting in a series of actions to correct the problem. However, if the firm's value system is based on the quick-fix mentality of a repair culture, nothing that you can easily do will change it to one focused on equipment reliability. Similarly, if the role models for the organization are not focused on reliability and the work needed to promote this behavior throughout the organization, then there is little you can do to introduce change.

Fortunately these two elements of culture are not mutually exclusive as they relate to work process. It would be out of place to have reliability-based values combined with role models promoting a repair-based process. It would equally be out of place to see reliability-focused role models working successfully with a repair-based set of values. Although either of these combinations is possible over the short term, a long-term working relationship is unlikely.

The other element of culture closely involved with work process is that of rituals and rites. The rituals carried out every day within the organization are the very work processes we have been describing. The rites are those events that the culture uses to reinforce the rituals.

For example, responding rapidly to an emotionally-based production problem in the plant is a ritual fostered by a repair-based work culture. The "pat on the back" for a "job well done," even if it may not have been necessary, is the rite that the culture uses to reinforce the ritual behavior.

Assuming that we have established reliability-based values and have in place reliability-focused role models, we still need to make changes to the rituals and supporting rites if we want to change the work flow (and supporting information flow) models to ones that drive the organization to improved reliability.

The remaining element of culture, the cultural infrastructure, also plays a role in the work process, even though it may not be as major as the other three elements. The cultural infrastructure is made up of people within the organization playing the roles of story tellers, keepers of the faith, and others who comprise this very complex system. Changes to the work process throw the cultural infrastructure into disarray. The stories no longer apply, the way things are done are different, and the information flow about what is going on changes. Once again, you need to know who occupies the key positions in the cultural infrastructure, then work toward getting them to support the change. Think of the value gained from such support!

10.9 Process Analysis

To change a work process, you need four things: 1) a vision, 2) a set of values that describes what the work process will need to support, 3) dissatisfaction with the current process, and 4) a plan. Assuming that your vision for the company is to have reliability-based maintenance, your values are reliability-based, but your work process is not (therefore leading to dissatisfaction), you have some serious work to do. The fourth part of the equation, a plan with the next steps, will be described in section 10.11.

The question now becomes how you analyze the current process to develop what is often called the new "to be" model. Developing and then implementing this model will enable your work process to align itself with your vision and values, overcoming the dissatisfaction with what is currently in place. This type of model can be developed using one of two approaches.

The first approach is to create an "as is" model. This model reflects what exists today. Once it is completed, you can then build the "to be" model, which shows what you want the process to look like when you have achieved your goal. Use the vision and the plant goals to develop the "to be" model. The team can then identify the gaps between the two models, recommending the changes necessary to bridge them.

The second approach ignores the current process. Instead, you start with a blank sheet of paper or an empty board. Create a process that reflects your vision. Do not be confined to the process used before. Because the starting point is so different, this approach usually leads to a very different result than the first approach. It allows more creativity. However, it is more difficult and time-consuming. In the end, you still need to identify the gaps and make changes to eliminate them. The difference is that the first approach is anchored in the past, the second approach in the future. For true improvement, I recommend that you use the latter approach for the analysis

You also need to be extremely careful in making up the team that you will charge with the task. All too frequently when the term reliability is mentioned, the immediate belief is that this field is the sole property of the reliability / maintenance department. This is not true. When building the team, make certain that production, maintenance, and reliability are all included as equal players. Failure to do this will leave you with a solution that, by virtue of not including all of the groups, will be unaccepted and fail.

10.10 How Do You Change the Process?

Once you have completed the design of the new process, there are essentially two ways to make the change. First you can take the step-wise approach, changing a little at a time. Some people believe this approach allows the organization to adapt to the new way of working and will lead to success over the long haul. I believe that this is incorrect. Step-wise change allows the new to be brought in while, at the same time, keeping the old in place. It keeps the old process in place as a sort of crutch. However, this crutch is exactly what the old culture wants because, if there is a problem, "we can always go back to how things were done around here in the past." Such thinking can be very detrimental to a new process and, with a strongly-resistant cultural infrastructure, can lead to its ultimate downfall.

I propose a different approach, based on a strategy used by 16th century conquistador Hernando Cortez. When Cortez and his army invaded Mexico to extract its wealth, they encountered early problems; many were ready to abandon the effort. One evening while everyone slept, Cortez burned all of the ships, making it impossible to go back. I suggest you take this approach with your change process. Do whatever is necessary to make a return to the old conditions impossible for the organization.

A reliability-based example with a similar approach involves a project of mine that involved redesigning a maintenance work process. The old process relied on maintenance foremen in each area reacting to the day-to-day needs of their production counterparts. The planners were also assigned to the areas and essentially acted in the role of materials coordinators.

Our task was to design a process that was planning- and scheduling-focused. To make this change, we went from an area model to a different model in which the foremen were centrally dispatched based on the weekly maintenance schedule. Because the area foremen were now centrally dispatched, production personnel had to go through the planner to get their work on the weekly schedule so that it could be executed. By eliminating the ability of the production personnel to bypass planning, we effectively made going back to the old way of doing business impossible.

The kinds of steps described thus far represent the short-term aspect of work process change They illustrate how you make the initial move away from the old process and implement the new. However, a second equally important aspect of work process change addresses long-term or sustainability issues. Unless you have an ongoing effort to maintain what you have changed, the culture will work very hard to restore the status quo.

The pitfall we typically encounter in process change is that we treat it like a work project; too often we assume that once the process change is rolled out, we are finished. This mistake is probably one of the biggest you can make! The truth is that once the change has been rolled out, you have just begun. Even if you have burned all of the ships the culture and those involved will find ways to slowly bring back the old way of doing things. This response is very detrimental if what you have eliminated is a reactive, repair-based maintenance culture because that is the last thing you want to reinstate.

10.11 Steps in a Process Change

I have been involved in many work process change efforts throughout my career. Some have been successful to varying degrees, some have been implemented with no apparent effect, and some have outright failed. Those reading this book have undoubtedly had similar experiences either being part of the process or being the recipient of a work process change designed by someone else.

Even those who succeeded probably did not bring about the level of change that was expected or hoped for by their sponsors. They may have created various rationalizations for why they did not realize their goals, yet never looked for the real reason. If they had looked, they would have discovered that, in every case, the culture is what got in the way of the successful work process change. The steps I recommend for conducting a work process redesign may not be as familiar as those that have been tried in your past. Nevertheless, consider them carefully. When necessary, review the chapters about the four elements of culture. The description of the steps is aimed at creating a reliability-focused work culture. However, this process will work on any work flow redesign effort you may choose.

Step 1 – Recognize that change is needed.

Every effort of this sort requires a recognition that change is needed. Recall that one of the three requirements for change to be successful (see Figure 3-1) is dissatisfaction with the old way of doing things. Often this dissatisfaction exists at the senior levels of the organization, but not where the rubber meets the road – at the middle- and front-line levels. Work process change will not be successful unless everyone has the same level of dissatisfaction. Thus, a profitable company that uses a reactive form of maintenance may have a more difficult time implementing a reliability process than a company on the verge of going out of business. Before you move from this step to those that follow, assure yourself that the same level of dissatisfaction exists throughout the organization, not just with those at the top.

Step 2 – Determine who needs to be involved in building the new process.

Often when a company decides it is going to make a work process change, the senior managers meet to lay out what the new process will look like. This approach is a huge mistake. A process designed in this fashion will fail, if not right away, then certainly over time.

Determining who will help develop the new process is a very important step. You will want a cross section of the functions within the maintenance department, but do not stop there. Your customer – production – as well as those who feed into the process need to be included. Therefore, your team should include representation from other departments. Finally, think about those who influence the cultural infrastructure – the "keepers of the faith." If you can get those who act in the mentoring role for the organization on board with the new process, then they will help sell it behind the scenes.

Selecting who will be involved does not necessarily stop when the team has been formed. You also need a champion at the correct level within the organization. This person must be someone to whom all of the groups report. For example, you can use the site manager if the site is going to implement the process. At a higher level, it could be a senior vice president if the process is going to be implemented across multiple sites. The champion can not be the production or maintenance manager because if one of them champions the effort, but the other does not agree, the process change will fail. However, if their collective manager is the champion, they will be aligned through the champion's influence.

Step 3 – Be clear on the values you are trying to achieve.

The next step is for the team to be clear regarding their organization's values. They must clearly identify what they hope to achieve and whether everyone involved with the initiative understand it to be the same thing. This step may seem simple. But each group starts with a different perspective of the work; time is needed for them to all share the same view.

Consider the statement "Design a process to keep the equipment running so that production can fulfill the requirements of the market." From production's perspective, this statement could lead to the development of a repair-based maintenance process. For maintenance, the process redesign may be somewhat different. Their concept of keeping equipment running may be tied to reliability efforts that actually keep equipment from ever breaking down. The appropriate time and effort must be spent to identify the values that we want the process to instill in the entire organization, not just one department.

Step 4 – Use the Goal Achievement Model.

Once you reach agreement on the values, your next step is to develop a Goal Achievement Model for the effort. This will allow the team to take the values (and long term vision) and develop specific goals, initiatives, and activities associated with a successful work process redesign. Developing the model will also vastly improve communication within the organization because it allows everyone to see how their individual efforts contribute to the overall success of the change. The Goal Achievement Model is discussed in Chapter 3.

Step 5 - Understand the current set of rituals and reinforcing rites.

Earlier in this chapter, I advocated preparing a redesign without spending an inordinate amount of time worrying about the old process. Even under this approach, you can not lose sight of either the current rituals (work processes) or the rites that are in place supporting them. Recognize that your new process will not only change how things are done, but will also change the way people are rewarded and their work within the old process reinforced. The culture does not take this without a fight. Consequently you must understand it and address it.

For example, consider a repair-based work culture where everything is an emergency; any work not handled during the day is then completed on overtime. In this scenario, a maintenance organization could have overtime as high as 20-to-30 percent. The existing ritual is the repair-based process; for the mechanics, the rites include large pay checks based on high overtime. Suppose the new process, designed to be reliability-based, will significantly reduce the overtime. How accepting do you think the culture will be of this change?

Step 6 – Conduct the work process design meetings.

This step is where the process is designed. Several aspects of this step need to be recognized at the outset. First, most plants are staffed for day-to-day operation. Getting people away from their normal job to work on this type of effort is difficult. Therefore, the champion needs to be the manager of all the affected organizations. Only then can the champion mandate that those on the team will be made available.

Second, this process does not involve only one or two meetings. Furthermore, the meetings will not be short and, for best results, the work should not be broken up into many short meetings, but rather a dedicated continuous process.

Third, the team needs a leader. Every group needs one and care should be taken in the selection process. The team needs someone who understands the business, is totally committed to the reliability-focused outcome, and has listening skills so that everyone can contribute. The worst possible leaders are ones who bully the team to get what they want and neither listen nor include the input of the other members.

Fourth, experts in the reliability process are needed. If you are trying to change from a repair-based culture to one focused on reliability, the expertise for the new process may not reside in-house. A consultant may be needed. There are many ways that a consultant can be utilized. Refer to my other book *Successfully Managing Change in Organizations: A Users Guide*, Chapter 8, for more details.

Last, the team needs process facilitation. Facilitators are people who keep the meetings on track and help the team work together to accomplish their task. They are not brought into the team to develop the end product. Nevertheless, their role is very valuable because they allow the team members to concentrate on the work.

Step 7 – Validate the work.

You can not build a team large enough to get input from everyone who will be affected by the outcome. However, if you have included individuals from all of the affected functions, you are certainly on the right track. This type of team has great value. When you reach milestones in the design effort, take a break, allowing these people to go back and validate the team's work with their peers. The input that they bring back to the group will be valuable and prevent the process from hitting barriers when rolled out.

Step 8 – Communicate.

The entire organization will undoubtedly know what is going on in your team. After all, many key players in their groups will be away from their normal jobs while they work on the redesign team. The culture needs to know what is going on so that it can take the time to think about the changes and, in many cases, expand the comfort zone to accommodate them. Not communicating frequently or with sufficient detail are serious errors. Without timely, detailed information, the cultural infrastructure – gossips, whisperers, story tellers – will have a field day spreading its own information. Unfortunately, that information is likely be inaccurate and raise concerns throughout the organization about the changes. You have a choice: tell the organization what is going on through constant communication or spend even more time later on damage control.

Step 9 – Training

As the process redesign becomes finalized, you and your team need to spend time designing the training program. Many well-developed work process efforts have failed because the training was poorly designed, inadequate in content, or poorly delivered. Many firms do not recognize how difficult and how important to success a truly good training program is; as a result, they do not give training the attention it requires.

My suggestion is that you hire a firm that specializes in the development of training programs. Although there may be employees within your company with the skills to develop these programs, they probably don't have time to dedicate themselves to it.

Two other advantages of hiring a firm to help with this effort are 1) their output will look professional and 2) the program will be designed so that it can be upgraded as the process evolves. The former is important for promoting the new design, the latter for sustainability.

Step 10 – Care for and feed the process (sustainability).

By the time you have reached this step, you have redesigned your process and implemented it in your organization. At this point, the team usually has a group dinner, everyone is recognized for their efforts, and all return to their regular jobs. The project is complete.

However, this is far from the case. If the change effort is approached in this manner, I guarantee that the best day for your work process will be the day it it rolled out. It will go downhill from there.

The reason is that you are working with a process, not a project. A process needs continuous care and feeding if it is to survive and add value to your organization. The process can never be allowed to stagnate. It must continuously evolve, as we'll see again in Chapter 12 where I discuss group learning. In addition, training can never stop. New employees enter your workforce all of the time and others change jobs within the organization. They all need training in order to sustain the new process.

To prevent these problems, you must keep the team intact. Although you may not need to meet constantly, you should meet periodically to review the health of the new process. One way to assess this health is to audit all or part of the process each time you meet. These audits are not done by sitting in a meeting room, but by actually going into the plant and examining the process where it is actually being used. In this way, process gaps can be identified and corrected before they get into serous problems. This type of audit, if done correctly, puts the audit team into contact with the cultural infrastructure, allowing you to assess what is happening behind the scenes.

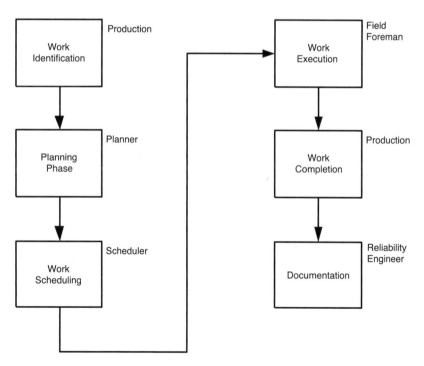

Figure 10-5 Simple Process Showing Roles and Tasks

10.12 Tools To Help with the Change

A difficult part of any work process change initiative is representing the new work flow in a way that can easily be understood by those affected by the change. Three tools will help you represent the work flow. Each has benefits and detriments. Choose the one that makes explaining the work flow process easier for you and for those who are going to implement it. Each tool may work well under different sets of circumstances.

I will explain each tool using a simplified work planning, scheduling, and execution process. Granted that a true work flow process will be far more difficult, but once you see how each of these are created, you will be able to take the leap from the simplified version to the complex.

Work Flow Diagram

Figure 10-5 shows a simple work flow diagram in which each block represents a step in the process. As you can see, the process begins with identifying the work and ends with closing out the work order. Next to each block I have listed the job that has primary responsibility for that part of the process.

The positive aspect of this approach is its simplicity. People can easily see how the work flows though the organization and who needs to act at each step of the way. What this approach does not show are the inter-relationships that each block has with other groups that do not have primary responsibility.

Referring back to Figure 10-5, the work flow diagram is where you can add in the information flow touch points if they are needed either for your analysis or to describe how your software systems tie to the work process. The two remaining types of work flow representations do not lend themselves very well to showing work and information flow on the same drawing.

RACI Diagram

The RACI diagram shown in Figure 10-6 addresses the weakness of the simple work process flow. RACI stands for Responsible, Accountable, Consulted, and Informed. The definitions of these four terms are as follows:

R= Responsible: The person participates in the completion of the task, adds appropriate expertise as required. There can be more than one "R" for each task.

Work Process RACI Diagram

Task	Production	Planning	Scheduling	Execution	Reliability
Work Identification	A			R	
Planning Phase	I	A		R	R
Work Scheduling	R	I	A		C
Work Execution	R	I	C	A	R
Work Completion	A	I	C	R	I
Documentation				R	A

Figure 10-6 RACI model

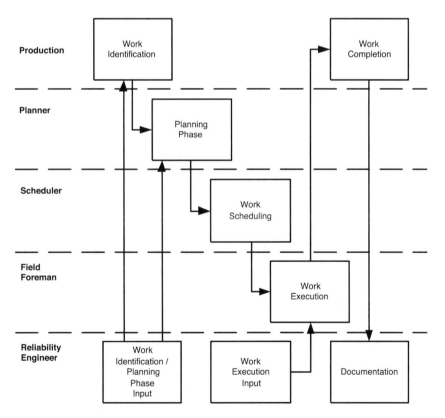

Figure 10-7 Complex work flow

A= Accountable: The person has sole accountability for completing the task; however, the task can be completed by one or more individuals assigned by the accountable person. Note that there can be only one "A" for each task.

C= Consulted: This individual(s) must approve or be consulted prior to the task being completed. There can be more than one "C" per   task.

I= Informed: This individual(s) is informed of the progress in com pleting the task. There can be more than one "I" per task.

The RACI diagram allows you to identify not only the person who has sole responsibility for a task, but also those who have ancillary involvement. What it doesn't provide is the flow of the work, as was depicted in Figure 10-5.

Work Flow / RACI Combination
Figure 10-7 solves the problems of the simplified work flow diagram and the RACI diagram.

As you can see in this diagram, we have taken the flow diagram that describes how the work flows through the organization and added on the left column the jobs that are involved in the process. In this way, as you track the flow of work, you can also see who has other forms of involvement with its various stages. This figure shows the involvement of the reliability engineer. The RACI diagram indicates the others who are consulted or informed; they can easily be noted on this diagram.

10.13 Process Change Results

Changing a work process from one that is repair-based to one that focuses on reliability of the plant assets is very difficult and time consuming. The change is usually managed incorrectly, leaving in its wake frustration and little value added for the firm. How it is implemented and how it takes the organization's culture into consideration can result in success.

The element of work process focuses on how the organization functions. A classic statement observes that "form follows function.' We have just discussed the function. In the next chapter on structure, we will discuss the form.

Structure

11.1 Introduction to Structure

Organizations can be structured in many different ways. Each way can provide value to the organization provided that the structure is the correct one for what you are trying to accomplish. You have probably been exposed already to many and varied organizational structures that exist within the arena of reliability and maintenance. A good structure can provide valuable support to the process of organizational change and the underlying culture. Conversely, an incorrect structure can seriously impair the ability of an organization to achieve its reliability goals.

This chapter will introduce the various types of organizational structures specific to reliability and maintenance. In addition, this chapter will show you how each of these structures can support or hinder the drive to improved reliability, specifically at the work culture level. Recognize that a good set of goals and initiatives, coupled with the best thought-out structure, can still fail if the organization's culture (and what you wish the culture to become) is not taken into serious consideration during the design phase of the effort. This task will be undertaken by relating this element of change to the four elements of culture.

11.2 What Is Organizational Structure?

When you first think about organizational structure, your organization chart comes to mind. This is the chart that uses groupings of boxes connected with lines to indicate where people fit within the organization. But structure is far more than boxes and lines on a piece of paper. It is the skeleton upon which your firm will create its overall work culture as well as the maintenance and reliability work culture. If the structure is created correctly and linked with the other seven elements of change, then change is highly likely to succeed. Conversely, if the structure is created using an incorrect model, something less than a successful cultural shift is far more likely as the outcome.

Structure can be defined as:a framework of roles, responsibilities, and reporting relationships that characterize how a group of individuals will work together to achieve a common purpose.

As you can see, this definition extends far beyond just boxes and lines. Structure is about what people do, how they do it, and how these efforts interrelate to accomplish a purpose. This definition and what it implies in the development of the proper structure are even more important in the arena of reliability. The reason is that most of us work in reactive work environments; altering the focus of the business means altering the structure of the firm as well. At the minimum, this change needs to occur at the plant level, and very possibly at corporate as well.

For our discussion, we need to understand two additional terms that apply to maintenance and reliability: repair-based and reliability-based work strategies. These terms will allow us to continue the discussion of structure from a reliability standpoint.

Repair-Based Strategy

A repair-based strategy is seen in organizations that have a "fix it now" attitude. The structure is built around the belief that, when equipment breaks, it is maintenance's job to restore it to service as soon as possible. This strategy usually has associated with it a high cost for maintenance service, a very short-term focus on the work effort, little planning, maintenance crews working at the direction of the production department, a high rate of emergency work (real or imagined), and very little of the failure analysis that could eliminate problems in the future.

Imagine it is a Friday afternoon and a major pump in the manufacturing process fails. The first thing production does is call maintenance to tell them that they have an emergency job requiring work around the clock throughout the weekend to insure that the pump is returned to service as soon as possible. The fact that production has a viable spare in operation is not at issue. What if the spare fails?

As a result of the call, maintenance works around the clock and repairs the pump by Sunday afternoon. To accomplish this, they had people working overtime who were not familiar with the equipment. They had to reuse some parts because new spare parts were not available until Monday. In addition, they had no time to analyze the failure. Still, the mission was accomplished.

What do you think management found on Monday? The spare, which was operating properly on Friday, was still operating on Monday.

Production did not suffer. In fact, production was reluctant to start up the primary pump because the spare was performing as required. The job was well done by maintenance, but neither the job nor the extra expense was needed.

Reliability-Based Strategy

A reliability-based strategy is far different. With this kind of strategy in place, the organization is structured to support the belief that things do not fail without warning. The organization also develops reliability-based work plans that proactively take corrective action to prevent failure.

This type of strategy is usually characterized by equipment ownership by production, high levels of performance monitoring to identify equipment that is beginning to fail, job planning, and the design of reliability into the work.

Let us go back to our previous example. The Friday afternoon call about the failed pump would not be as serious because, through the monitoring process, the pump would still be in the early stages of failure. When contacted, the reliability group would have production put the spare into service. After checking out the failed pump and determining that it was in good enough shape to be repaired, a work ticket would be put into the maintenance system. There would not be any overtime nor frantic repair efforts. During the following week, the job would be planned; the new seal designed to improve performance would be included in the work package. When the work was executed, an engineer would be present to conduct a detailed failure analysis.

Comparing Strategies

These two types of maintenance / reliability strategies have a great deal of impact on organizational structure. Look at organizational structures throughout various industries, we see that they have been built in direct support of the company's vision and values related to maintenance and reliability. For those operating with a repair-based strategy, you would find maintenance work crews and even the entire department working for production, much more reactive work, more emergencies, and very little planning. In this type of environment, the reliability engineers (if they exist at all) are there to keep the plant running. They only have a day-to-day work focus. For those operating with a reliability strategy the roles, responsibilities, and work efforts of the maintenance and reliability functions would be very different. When designing an organizational structure, keep in mind 1) form (the organizational structure) follows the function (work process) of the organization that it is designed to support and 2) the function follows the vision.

11.3 Reliability-Based Structures

Before we begin a discussion of organizational structure focused on reliability, we need to understand a very important concept about reliability at the plant level.

Reliability is not a department. It is a way of thinking about the plant equipment – a mind set. As a result, production, maintenance, and reliability engineering have equally important and overlapping roles and responsibilities, all to ensure that a high level of plant reliability is maintained.

Furthermore, to be successful, these functions must interact daily in many different ways. As a result, how you build your structure defines this interaction and reinforces the overriding work culture that you are trying to promote.

Let us clearly define the three critical components of a reliability structure. We can then discuss the types of reliability-based structures in the next section.

Production

The production group operates and ultimately owns the equipment that takes raw materials and converts them to salable products. Consider your car: What is it that you want your car to deliver every time you turn the key? You want it to start and safely take you to your destination, in other words, reliable operation. Would you knowingly run it with the oil light lit or with bald tires? Probably not because you own that car and need it to operate reliably each time you turn it on.

Maintenance

The maintenance group does proactive repair and, if necessary, fixes things that break. They change the oil and rotate the tires. In addition, they identify and repair problems when you bring them to their attention.

Reliability Engineering

The engineers (also called maintenance engineers) form the group that identifies longer-term problems and designs ways to correct them.

They typically do not work on day-to-day issues like maintenance. Instead, they are focused on longer-term repairs. Continuing with our car analogy, they are the ones who would design an improved suspension for the next model release. They also are the ones who address chronic problems and provide solutions. Whereas the maintenance group would repair defective parts, the reliability engineers would focus on preparing a new design so that future manufacturing would not have defects..

Each group has a direct impact on reliability, but in different ways. In addition, each needs to have interaction with the others to perform their assigned task properly. You must consider these two factors when designing a reliability-based work structure.

11.4 Reliability-Based Structural Components

Each of the three functions that we discussed in the last section — production, maintenance, and reliability — has different roles that interact and overlap on a daily basis. Although they do interact, you should not try to combine them into a single function. For example, you can not successfully give a reliability engineer both reliability and maintenance duties. Similarly it is difficult to have a production supervisor handle maintenance tasks. These are separate skills requiring a different focus and way of thinking. Furthermore, each of these jobs has pressures and demands. If they are combined, one or the other ultimately suffers.

In order to build a structure that permits separate job roles and responsibilities while still allowing for overlap of the functions, two different types of organizations have emerged. The first, a functional centric (centered) organization, is designed with the individual job functions as their focus. The second is designed with more focus on the work process and as result is called the process centric organization.

Functional Centric Organizations

In the functional centric organization, the three functions we have identified usually work independently of one another while at the same time working together to keep the plant running. This structure creates a problem for those who hold the reliability and maintenance positions. If the managers of the reliability and maintenance functions and the production manager are not in goal alignment, then those in the functional departments get pulled in two directions. Their functional manager may want them working on a strategic project that requires a considerable amount of their time. At the same time the production team may be holding a meeting on a critical production problem that requires their input. Obviously this creates conflict and disharmony. Not only does production suffer, but so do those caught up in the pull between two very different and equally important responsibilities.

Process Centric Organizations

In the process centric organization, the functions usually work in close proximity with each other. They are very day-to-day focused, with production often dictating the primary focus. The problem here is exactly the opposite of the functional centric organization. Because the focus is more day-to-day, reliability and maintenance functional managers feel cheated, unable to work on the more strategic projects that will benefit their functions and the company over the long term.

As you can see, each type of organization has pros and cons in its design. Figure 11-1 shows both types and their positive and negative aspects.

The Hybrid Organization

So what is the answer? Neither of the two organizational focuses

	Functional Centric Structure	Process Centric Structure
Focus	• Individual organizational functions	• Manufacturing process that the functions support
Positive	• Ability to work outside of the day-to-day efforts on more strategic projects. • Close proximity to others in the function to promote interaction and mentoring of new / less experienced personnel.	• Team approach focusedon getting and keeping demand of production satisfied. • Close daily interaction. The team can quickly andcollectively address issues.
Negative	• More attention to function and peers vs. process. Process could suffer. • Interrelationship problems between functions as group members find it difficult to of functional skills.allocate their time.	• More focused on day-to-day production. The tactical is addressed at the expense of the strategic issues. • Lack or reduced mentoring • Lack of consistent approach due to a weaker functional connection.

Figure 11-1 Functional vs. process centric table

appears to be optimal. Figure 11-2 illustrates a hybrid structure that serves the combined needs of both organizational designs. In this figure, the y-axis depicts the percentage of the organization's focus on strategic or tactical work from 0 to 100 percent. The lines labeled Tactical and Strategic show the approximate percent of the functional and process models in each. For example, the functional centric organization has a higher percent of strategic focus and a lower percentage of tactical focus whereas the process centric model has the opposite balance. Because neither is optimum, I have shown a new model which is labeled as a hybrid of the two. Neither wholly tactical nor wholly strategic, it is designed to address the needs of both. As you can see from Figure 11-2, the hybrid model addresses the needs of both structures. The use of this model allows you to gain the positive aspects of both types while at the same time eliminating, or at least minimizing, the negative impacts.

11.5 Creation of the Hybrid Structure

What we are trying to achieve is a structure that will provide us both the functional and process benefits, one that will initiate and support our goal of plant reliability. However, one of our assumptions is that an individual can not do both well; the structure must take this into con-

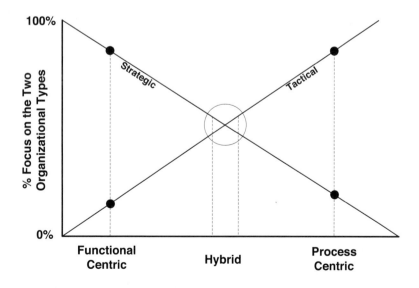

Figure 11-2 Functional vs. process centric diagram

sideration. Therefore, it will have functional and process centricity by creating two separate but interlocked groups.

Understand that the structural model I am offering should be used only as a starting point for your structural design Every company does things differently. Therefore, the model may need to be modified to support your own specific requirements. However, the concept is sound and should be used as the foundation for the hybrid structure that you create.

The Functional Group

To a large degree, the functional group needs to be removed from the day-to-day work process. Although they will be working on projects to improve day-to-day production, they have longer term commitments that require an almost dedicated focus. Other than interacting with the tactical group to recommend improvements and to seek input for their ideas, the functional group works independently. This independence gives them time to think and plan for the strategic efforts that need to be implemented to improve reliability and add value to the business. The people in these groups are usually senior members of their organizations. They have acquired the practical experience needed to work on these efforts.

The Process Group

The process group is involved in the day-to-day workings of the business. Their job is to keep the plant running. As part of this effort, they own the equipment and must pay careful attention to its reliability. This part includes monitoring the equipment and repairing it as necessary. The other part is their connection with the functional side. Because they are monitoring the equipment, they can predict impending failure well before it happens. This ability allows planning for the repair event and including the functional group in deciding how the repairs can add improved reliability to the asset.

Individuals occupying the process side of the structure are often the less experienced representatives of their functions. Working in the line organization, the real-time experience they obtain will make them even more valuable as they move into the functional area.

Group Interaction

In each of the two groups above, I have referred to both the concept of independence of the functions, but also the need to interact. This need follows from Figure 11-2. Even though each type has a predominant focus – either tactical or strategic – there is some degree of the other type present. This interaction provides a communication path so that functional ideas can be validated by the process and process activities can be monitored and mentored by those within their functions.

Figure 11-3 graphically shows the functional and process relationships. The functions reside in their own circles labeled P, M, and R. They are largely independent in nature, working longer-term strategic initiatives. They have a loose connection as identified by the outer ring labeled functional groups. Each of these independent rings overlap indicating that none can not exist in a vacuum. I have indicated the overlapping functions that are addressed. For example, P-M shows the overlapping of the production and maintenance parts of the organization. The three types of overlap (P-M, M-R, and R-P) are indicated by the middle circle labeled interaction. Finally, the center shows where the three functions all overlap with one another. This area, the process part of the organization, is represented by the circle enclosing P-R-M. A typical application of this type of structure is the use of area or process-related business teams.

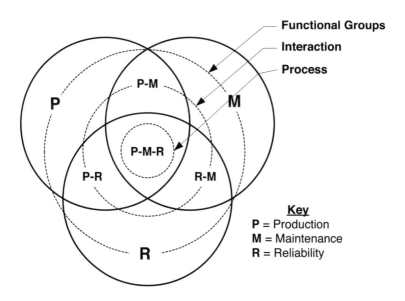

Figure 11-3 Functional vs. process overlap diagram

11.6 Organizational Geography

Another aspect of organizational structure is your office's location relative to your organization. The type of office to which you are assigned is also important. Office location sets the tone for the work process; it has the potential to reinforce the type of culture that you are trying to create. In addition, it can be beneficial or detrimental to the organizational structure that you are implementing.

Typically people are grouped in functional- or process-centric units. This grouping is often the result of structural evolution and is not done by design with a long-term vision in mind. As an organization begins to grow, the hiring group tends to locate the new employees in close proximity, often to help them adjust to their new jobs. In addition, the more senior employees can more easily oversee the work of their juniors.

As time goes by, the group grows and so does the office space they occupy. This geographical proximity works in favor of perpetuating the culture. The cultural infrastructure is anchored by the "keepers of the faith" who work in a mentoring capacity, getting the newer employees to understand how things are done. The close proximity also allows the mentors to see deviations and take immediate corrective action. Thus, the status quo of the culture is maintained, whether the structure is process or functionally centric. Let us take a look at the pros and cons of these two location strategies.

Process-Focused Location

Pro: Those located centrally to the process are entirely process focused. They pay close attention to what it takes day-to-day to keep the plant running. They feel strong allegiance to the individuals running the process and work diligently to support them.

Con: With their day-to-day focus, they do not have a large amount of time for strategic efforts because it takes them away from their primary function. In addition, they often focus on the process at the expense of their functional leadership. Because they are isolated from others in their function, it is difficult to be mentored or to upgrade their skills.

Functionally-Focused Location

Pro: Those located within their function have the time and ability to focus their efforts on strategic work initiatives. This focus is supported because they have peers surrounding them who can act as sounding boards and provide input. Because functional structures are usually located away from the day-to-day effort, they can work without interruption.

Con: Functional experts are often needed to work on day-to-day problems. The value that they add is that they are usually the most experienced people in the organization. However, being completely isolated from the line organization leaves them isolated and often not consulted. This can get them angry if they believe they are not valued. In addition, their isolation creates hostility with those on the process side if they believe the functional experts do not want to "get their hands dirty."

The problem with organizational geography is the same that exists between the process and functional structures. Each structure has its benefits, but neither can not work independently. The overlap described in Figure 11-3 must be a factor of where people sit when you build the structure. Thus, where people sit has a great deal of importance when you decide to restructure an organization. This importance is most obvious when you try to implement a culture that is based on the principles of reliability.

Years ago I worked in an organization that centrally dispatched the maintenance crews to the work based on job priority. In this model, each planner focused on a specific area of the plant. However, the foremen who were dispatched from a central area could work in one process area today and an entirely different one tomorrow. This same approach held for the mechanics as well. At the time, we were still very reactive in how we conducted our business. In fact, the work process change that created central dispatch was designed to break the reactive model, making us more focused on planning, scheduling, and a reliability-focused work effort.

What we had created by moving to a central dispatch process was a functional structure. We coupled it with a functional geography – all of the foremen sat together in the central area awaiting assignments for their crews. Although the foremen were able to interact and help each other, the process side suffered. Because the foremen could wind up working at any location within the plant, those operating the process were unable to have continuity for either the supervision or the craftsmen. In addition, they found it difficult to speak with the foremen because their offices were not in the process areas. Over time, this problem was recognized as being detrimental to the effort. Subsequently, the foremen were reassigned to the process areas.

In another example, the planners were assigned to the production areas. In this design, they lost their planning focus. Because the plant was highly reactive over time, they became nothing more than material expediters for the work crews in the areas.

The issue of location is important because it can either support whatever structure you are trying to create or tear down a structure to which it runs counterproductive.

11.7 Structural Pitfalls

Thinking about an organization's structure in terms of a successful cultural shift is not all that common of an approach when changing the structure of an organization. Typically organizational structure is based on other factors.

We want to make improvements in the process, so why not restructure?
Changes take place when we determine that improvements are needed in the process, then decide that one way to accomplish this is to change the structure. I believe that this is the most frequent type of organizational shift. The motivation of those rearrangements is usually well founded. Yet too often we believe that rearranging the boxes, and often the lines, without anything more than a loose concept of a desired end state will make a difference. Making changes in this fashion often makes things worse and harder to correct latter.

Many years ago I worked for a manager who believed that he could improve reliability of the plant by changing the organizational structure from one that was decentralized (foremen and work crews assigned to the specific process areas) to one that was centralized. In this new model, all of the work crews would be dispatched by priority. Theoretically the idea was sound and, if approached correctly, could work. However, all he did was rearrange the boxes without much thought to the resultant impact. What happened was that, in one massive change, he broke the culture apart. The values, which were focused on close relationships between maintenance and production, were broken. The role models that existed before the change were no longer in place to advance the company's reliability concepts. The rituals and their supporting rites, which were process centric, were broken and the cultural infrastructure was disrupted.

The effort met with strong resistance and ultimately failed, returning us to the area concept and providing us with a new manager. Although the effort eventually worked out in the end, the disruption and resulting problems caused by the change were not worth the price.

A change in management after which those newly installed in their positions change the structure to suit what they believe is a better way to manage.

In this model, new managers are hired to correct plant problems. For our example, these problems are related to reliability and maintenance. The new managers were presumably successful in their prior employment; otherwise, they would not have been hired. They were brought into the plant to make things better.

The mistake the managers make is believing that they need to reorganize to show everyone that the old way is gone and a new process (and managers) has arrived. To make matters worse, this reorganization usually takes place shortly after their arrival – long before they have had a chance to try to understand the culture and address its idiosyncrasies.

The result of implementing change in this manner is the same as in our previous example. Because the managers have not taken the culture into account, the culture will work to restore the status quo and ultimately win out. An additional problem in this model is that the new managers are unlikely to give up easily. After all, the new model that they are implementing made them successful in prior job assignments. As a result, conflict will arise between the managers, as they try to implement a new structure, and the cultural infrastructure working to maintain the status quo.

Adjustments to a structure to accommodate individuals and their specific skills.

Years ago I worked for a manager who had superior skills in production, maintenance, and reliability. Given his abilities, when we reorganized the management team, we consolidated the managers of production and maintenance into one position called production manager. My manager, qualified in both areas, was given the job and he executed it well. The problem with this form of structural change is that it is based on a single person's set of skills. When my manager moved on to another job, there was no one with the same set of skills to replace him. A very high-

powered manager was put in the position, but his skills were more production focused; over time, maintenance suffered. Several years later,
we went back to separate managers for each of the functions.

Recommendations from a process consultant

The other type of organizational change comes from the work of
process consultants. Many times when we hire consultants to help us
improve our work process, their deliverables include recommendations
to change the organization. This type of structural change has the most
chance of success because consultants understand the global nature of
process change and consider restructuring along with the other elements. Their recommendations usually include transition planning and
training so that the shift is accomplished with a focus on success.
Although potential pitfalls remain, at least direction is coming from
someone experienced in this type of process improvement.

11.8 The "How To Do It" of Structural Change

Restructuring takes place all of the time in business. We have talked
about the pitfalls of restructuring for the wrong reasons or with the
wrong focus. The real question is how do you do it correctly. My concept
is not the same as what you may have heard about how to restructure.
It takes a different approach and brings organizational culture into the
discussion. You can not restructure without seriously addressing the
cultural issues that already exist or the new cultural dynamics you will
be creating with your new organization.

Only two steps are identified here. If you address these before you
start the restructuring exercise, you will increase your chances for success tenfold. Once these steps are addressed, the other steps are the
same that you have used numerous times in the past. For additional
information regarding Structure, see Successfully Managing Change in
Organizations: A Users Guide – Chapter 11.

Know your vision and your values.

Start the restructuring effort by clearly knowing your vision of the
future as well as the value system that you wish to have in place to support it. Remember: A vision that is reactive (repair-based) will require
a structure far different from one that is reliability-focused. Building an
organizational structure that won't support the vision will create high

levels of frustration within your organization as well as with your customers.

Address the organization's culture.

Before you implement the change program, address the organization's culture, specifically the four elements of culture described in earlier chapters. Only then can you prepare the organization for the change that is coming.

The first element of culture – values — needs to be aligned with the new structure. If you currently have a reactive culture, but are trying to switch to one that is reliability-focused, the current cultural values are not in line with the new way you plan to do your business. In this case, reread the chapter on organizational values. Then put in place a plan to communicate; generate buy-in from the organization for the new work focus. Only when buy-in is achieved can structure be changed.

Rituals and rites are also extremely important to have in alignment before you alter the structure. You need to fully understand the current set of rituals and the rites that reinforce them. Suppose that the rituals are focused on the plant emergencies or problem of the day at the expense of planning and scheduling. The reinforcement that the maintenance people get for quick response (the rite) is a difficult thing to eliminate for both the maintenance and production personnel. Until you determine how to alter the rituals and their associated rites, however, structural change is not ready to take place. For example, think how frustrated the maintenance foremen and their crew will be if the new structure eliminates their rapid response and pat-on-the-back work process.

Role models also need to be addressed prior to the change in structure. In a reactive work environment, the role models are far different than those in a reliability focused one. In the current process, those engaged in the work look to their reactive roles models for guidance and support. These are the people who are looked up to and used as models for behavior. They are respected because they support the reactive process. What will happen if you restructure in a way that removes the influence of the current role models, replacing them with people who the current work process does not recognize as major players. Consequently, you need to have the right role models in place before restructuring takes place. Doing so can take time, but positioning yourself in this manner will provide you with a higher likelihood for success.

In addressing the culture before you change your structure, you must also carefully consider the cultural infrastructure. The symbols (measures) that are used to designate success, how you persuade the "keepers of the faith" to adopt a new way of working, and the communication structure you put in place to keep the organization informed before the change — all of these are important to your efforts. Addressing these aspects of the cultural infrastructure before changing the structure of your organization reduces the likelihood that the structural change effort will be derailed before you start.

11.9 Close

Whatever final structure you create to achieve plant reliability – process, functional, or hybrid — your ultimate goal is to increase the organization's focus on plant reliability. To accomplish this goal, you must focus first and foremost on culture and its primary components: the four elements of culture. If you fail to focus properly, you run the risk of a structural – cultural mismatch, creating a structure that will not be supported by the plant's culture. Over time, the culture always wins. Therefore, at the beginning, you need to recognize the importance of culture, understand it, address it in relationship to structure, and make your cultural shift before you make your organizational shift.

Group Learning

12.1 Introduction to Learning

This chapter is about group learning that is specifically related to organizational change, and even more specifically related to plant reliability. Every organization and its employees learn continuously. Although you and your organization may not learn the correct things nor apply what you have learned, nevertheless, learning goes on all of the time.

The questions we will address in this chapter are

1) What is group learning?

2) What constitutes a learning organization?

3) How do we acquire knowledge?

4) How does group learning apply to reliability?

5) How is learning supported or impaired by the organization's culture?

6) What we can do about it?

In our daily lives, we learn something new everyday. It is important that we consider each of these new things that we learn, and then see how they are applied to the world in which we work. If we don't, there is a significant risk that the world will change and we will be left behind. Think of the things you have learned over your career—they have been applied towards making work easier and making you more productive? Computers are a simple example of learning that has affected everyone. I can still remember doing things long hand and the large amount of time and effort it took. I can also remember the difficult time and extended effort required to justify one PC in our entire organization. Times change; we must learn and apply what we have learned.

Even more important for the success of our companies, we must learn as a group, hence group learning. No one individual can easily make a change for the better. You may have learned something that can vastly improve the reliability of your equipment, but if that learning is not transferred to the entire group, it will never be implemented. Groups are what make up the organization and, ultimately, the company. If the group learns something new that will support their effectiveness and efficiency, and they are aligned in their approach, the result will be a powerful and successful change.

Years ago, I learned that reactive maintenance was not nearly as effective as work that was well planned and executed. However, many co-workers still embraced the old reactive way of working. At one point our expenses were extremely high largely as the result of poor reliability and our "break it – fix it" mode of operation. We were told that we would be either sold or shutdown within a matter of six months. With plant closure looming on the horizon, the organization was open to any ideas that would improve our performance and make us an attractive purchase. They were willing to learn new ideas, among which was planning and scheduling. We implemented these concepts and improved our performance. I can not say that this was the reason we survived, but our improved productivity and improved reliability certainly helped. As a team we learned; the power of implementing the new ideas was a significant factor in our purchase.

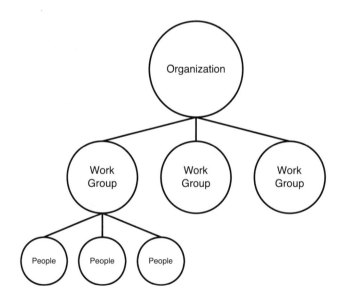

Figure 12-1 Organizational learning levels

12.2 Group Learning Defined

For an organization to make a successful change from its current state to one focused on reliability and proactive maintenance, it must have group learning as one of its key elements of change.

To better understand the concept of group leaning, we need to define this term. We then need to break the definition into its component parts, developing an equally deep understanding of each part.

> Group learning is the ability of individuals to acquire new knowledge, then use this knowledge in a group setting that either leads to aligned alternative courses of action or reinforces that the current course of action is correct.

Let us break this definition down so that we can discuss its significant parts and better understand what group learning for an organization is all about.

12.3 Levels of Learning and Alignment

Organizational learning takes place at three levels: the organizational level, the work group level and the individual level. Each is a subset of the prior group. The work group is a subset of the organization and the individual is a subset of the work group. Figure 12-1 illustrates these relationships.

As you can see, individuals are part of work groups. Although we learn at an individual level, the learning never fully takes root unless the idea is embraced by the team of which we are a part. Later we shall learn how the culture of the organization and the working groups (subcultures) influence learning.

The work group level is where learning can gain a foothold. However, this is not as simple as we often want it would be. Work groups are dynamic. People learn in different ways and draw conclusions that may be different from other members, even though they are presented with the same basic information. Suppose at a recent seminar you learned about a new technique for aligning rotating equipment. From this experience, you recognized that this new procedure could significantly improve the effectiveness of your plant's mechanics. However, when you introduced the concept to your work group, many members resisted the idea. Some believed that it would diminish their skills, others believed that the new process could only be handled by a few experts and ultimately cost them their jobs, and a few liked the idea and could see the benefits.

Thus, a simple new idea that would improve how work gets done met with various reactions from the group – most against the idea. This misalignment of what people understand from the information that you provide them can seriously derail even the best of ideas. It can be overcome by slowly introducing new ideas and handling them on a pilot basis. Then everyone has the ability to see the benefits and understand the impacts. When this approach is taken, successful implementation of a new idea is generally made easier because people have time to learn more about it and ultimately accept it.

If it is hard to achieve alignment in a group, imagine how hard it will be to take learning from one group and apply it across an organization.

12.4 How We Acquire Knowledge

We acquire new knowledge from both external and internal sources. Externally we encounter new information all of the time. We learn from participating in conferences and seminars, meeting with company representatives on industry committees, working with external consultants, attending training classes, and many others. Internally we learn from our constant interaction with other plant sites, other departments within our own site, and new hires who have a more objective view of what we do – at least for the first months of their new careers.

The problem we face is that we continually filter information. Learning new things is one step towards change. The harder step is being open enough to allow new ideas past our filters so that we can analyze the information and determine if it is applicable. We need to be open to the fact that someone else may have invented a better mouse trap. If we fail to turn off the filters, we are doing a disservice to our company by potentially missing out on new ways to improve the work. Filters of this type look and sound like:

  • That is not how we have done this work before.
  • We have been successful with the old method – why change?
  • We tried that before and it was a waste of time.
  • Our way of doing _____ is better.
  • That is a different industry – it won't work in ours.
  • That can't be right; our information says something different.

Many others could be added to the list. Each in its own right is a destructive force that can render the best new idea useless. Rather than dismissing new ideas, analyze their value. When there is value, use the idea to improve the business. When there is no value, dismiss the idea, but at least you will know that its potential was evaluated first. The learning organization can recognize the gaps within their own sphere of knowledge then create or acquire new knowledge to fill these gaps. Keep in mind that filters can also provide a useful purpose. If used constructively, they can provide a framework that will enable you to pre-screen information so that your efforts are focused. In that way, they help you to keep too many low-value ideas from overpowering the good ones.

12.5 How We Employ This Knowledge

In Chapter 8, I discussed feedback loops as a part of the learning process. Figure 12-2 shows a single loop where we set goals, act on the goals, and compare the results of our actions. Where there is a gap between our goal and the experienced outcome from the work, we identify it and feed it back into the work system. This causes us to readjust our work activities so that our next outcome more closely matches the desired goal.

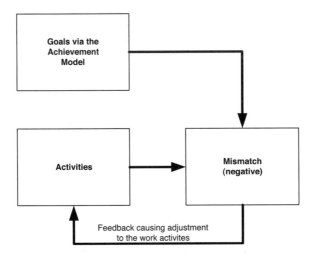

Figure 12-2 Type 1 learning

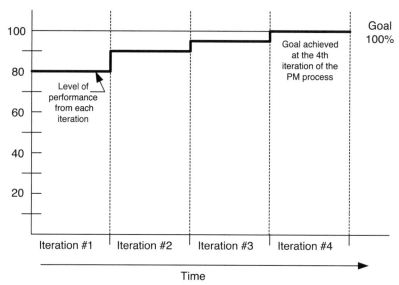

Figure 12-3 Type 1 – achieving the goal through repetitive iterations

Through continual application of the learning loop, we eventually close the gap and reach the goal.

At this point, we may want to readjust the goal. New goals that are more challenging that the previous ones raise the bar. Suppose we want 100 percent compliance with the preventive maintenance program (PM) so that PM tasks are completed as scheduled. After conducting our program, we measure the results. Assume that the first time we do this; our compliance is at 80 percent. This result is then fed back into the work process; changes are made that will affect the rate of compliance. Assume that the outcome of the second iteration is now 90 percent. Once again we review the work outcome and adjust the process until, at the fourth iteration, we reach the goal. Figure 12-3 demonstrates this process graphically.

The above single loop learning process is relatively simple to apply to single activities such as a PM program, but think how much more difficult it becomes when applied to the many and inter-related goals of the business.

Furthermore, the single loop learning process has a flaw. What if the goal against which you are comparing your activity outcomes is wrong? Not possible, you say! Consider organizations that have rapid repair or

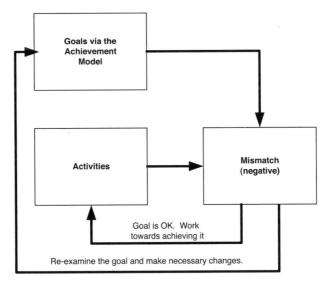

Figure 12-4 Type 2 learning

"break it – fix it' as their mode of operation. Their goals are centered on improvement to the repair process. These organizations have feedback loops telling them to strive to fix things faster.

We already know that in most instances this goal is expensive and inefficient. Therefore, what can be done to break this flawed feedback process? The answer is the employment of double loop learning as shown in Figure 12-4

In this model there is an additional process. Not only do we examine the outcome of our activities versus our goal and make needed adjustments, but periodically we also re-examine the goal and make adjustments here as well.

Let us go back to our PM example. Suppose that when reached the 90 percent compliance level, we noticed that the benefits from attempting to reach 100 percent were minimal when compared to the amount of effort, time, and money that would be required to achieve full compliance. Do we want to continue chasing this goal? In some circumstances, the answer may be no. Instead we need to re-examine the goal and make adjustments at a higher level. Our goal may need to be changed to 90 percent PM compliance (for example, not worrying about PM on some equipment) and 10 percent predictive maintenance. This brings into play a different feedback loop – not working to alter our activities, but to alter our goal.

Understanding and employing this process is often new, different, and difficult for an organization. However, as you learn you evolve; the goals that you set at the beginning need to be altered based on your learning. We will talk more about how this works when I address learning spirals. Realize, though, that if you always conduct business in the

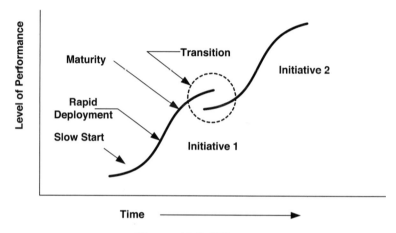

Figure 12-5 "S" curves

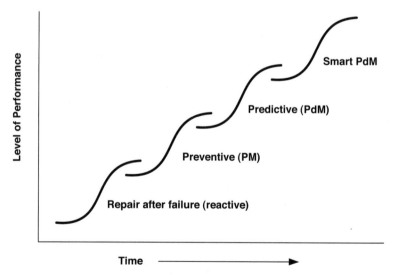

Figure 12-6 "S" curve example

same way, you can never expect to get anything other than the same results of the past.

The next question that needs to be addressed concerns when to re-examine the goals. The rule of thumb is that goals should be re-examined when the current level of activity is no longer generating incremental benefit. This level varies for different organizations. It is not usually tied to a calendar, even though companies seem to believe that goal setting is an annual event.

Goals, their review, and often their readjustment fit an S-curve model. Figure 12-5 shows how this model works. The x-axis represents time and the y-axis represents the level of performance from low to high. The activity that we are tracking against our goal starts off slowly. As time progresses, it is accepted. Then, as performance feedback is fed back into the process, performance begins to rapidly improve. The initiative reaches maturity, after which progress slows, often reaching a plateau where addition improvement is non-existent.

At this point, the second loop of the double loop learning model needs to be employed. As the existing goal is reviewed, we recognize that it is time to make a transition to a new goal that once again will add significant value to the company.

Figure 12-6 shows an S-curve for maintenance in some companies. In this model, the company begins in a reactive mode of operation. Over time it learns that this way of working is both inefficient and ineffective. It then shifts to a PM focus. After a while, this program also fails to deliver the value that the organization wants. Next, they shift to predictive maintenance and so on through continued feedback loops that alter the organization's goals.

12.6 Spiral Learning

The single and double loop learning models provide us with a valuable insight into the process of learning. Most important, the process of learning and using new ideas and tools is not linear. Learning does not proceed as a work project would in which we know what the outcome will be before we start the work. Learning is far different. As shown in Figure 12-7, it takes place in spirals that are similar to the single and double loop models.

In spiral learning, we identify a new concept or idea, plan our actions, execute according to our plan, and then review the outcome using the feedback loop models. Through the feedback loops, the review

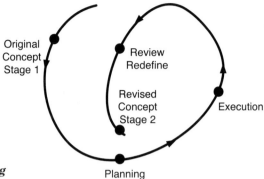

Figure 12-7 Spiral learning

will direct us to alter our activities or even our goals. When we start any loop, we often do not know what the second, third and future loops will look like until we get closer and have learned from our actions. This point is especially important for engineers and their managers who have been trained to think linearly. However, linear thinking does not work well in the world of change where activities, results, and future activities are non-linear. For a more detailed review of this subject, see Chapter 3 of *Successfully Managing Change in Organizations: A Users Guide*.

12.7 Learning and Blame

As we perform various activities in an attempt to achieve our goals, things often go wrong. The wrong procedure is applied to the work, the wrong part is installed, and the equipment is started up incorrectly, a whole host of other things that affect production and performance may also go wrong. A learning organization will approach these failures far differently then another type of organization, which I will call a blaming organization. In the learning organization, the leadership will ask "What can we learn about this problem so that we can avoid it ever happening again?" A blaming organization will search for the guilty, and then assign blame and punishment. In the learning organization, improvement is a definite possibility; in the blaming organization, a repeat of the problem is a certainty.

A learning organization will work hard to search out the root cause of the problem and, in turn by correcting the root cause, eliminate the potential for reoccurrence. This goal can never be accomplished if blame and punishment are part of the equation. The reason is that people will not tell the truth about the incident for fear of being blamed and punished. Why would mechanics or production operators in a blaming organization admit to making a mistake if the consequences could be avoided by playing dumb? Therefore, they say nothing and the real problem—that the wrong part was in the warehouse or that the procedure for start up was wrong—goes unnoticed. As a result, we do not learn from our failures.

A learning organization addresses each of the above problems differently. It provides amnesty for these individuals, guaranteeing that there will be no punishment if they reveal what happened during the investigation. In this way, learning organizations can get at the root cause of the problem and prevent reoccurrence. However, an amnesty policy sets up a conflict. How do we provide amnesty and still hold people accountable for their actions?

Two types of actions need to be considered if we are to address the question about amnesty: inadvertent actions and willful actions.

Inadvertent Actions

Inadvertent actions are steps that we did because we believed that they were correct. When inadvertent action causes failure, blaming the individual that caused the actual failure is useless. They are the last link in a long chain of failures that led up to the event. Take, for example, the wrong part in the warehouse. Is the person who installed it after taking the number from the bill of material to be blamed? No. The cause of the failure needs to be spread far across the organization. Part of the cause was the engineer who wrote down the wrong part, the purchasing agent who bought it, the vendor who should not have supplied it but did, the rotating equipment engineer who told the mechanic to install it, and finally the mechanic. Learning and, ultimately, corrective action are far better served by making certain that everyone understands that there is no blame and punishment for this type of failure. In this way, through amnesty we will learn the root cause of the problem and the wrong part will never be installed again.

Willful Actions

Willful actions are steps that we did even though we know at the time that what we were doing was wrong. This type of action is far different from inadvertent actions. In these cases people, are breaking what are referred to as the Laws of the Land. These actions do and should bring blame and punishment. For example, mechanics who bypass a plant safety rule, thereby exposing themselves and their co-workers to serious injury, have broken a Law of the Land; they deserve the blame and punishment that accompanies it.

The important thing for a learning organization to provide at the work site is a clear understanding of what the Laws of the Land are. In this way, people will easily be able to differentiate between inadvertent action and willful action. The amnesty rules can be applied to improve our learning from our failures.

12.8 Learning and the Eight Elements of Change

It is easy to see how learning can be applied to the other seven elements of change. All of these elements, along with the actions we take related to them, are tied to single loop and, quite often, double loop learning as we develop and improve upon our goals. When learning is not an active element of change, we are doomed to repeat our mistakes, never really understanding their cause. A quotation that immediately comes to mind: "If you do the same old things in the same old way, you will always get the same old results." Isn't this exactly what happens when we fail to learn and apply what we have learned?

All workers need to keep their eyes and minds open to new ideas and think about how these new ideas can be applied. The eight elements of change provide a framework in which to conduct this learning examination. For example:

Leadership – Identifying the lack of leadership skills in key individuals who you need to champion the change initiative, and then providing them with the training they require to be successful.

Work Process – Recognizing that a reactive maintenance work process is both inefficient and ineffective, then developing a proactive process to improve equipment reliability.

Structure – Identifying the fact that the current work structure may not support your longer-term reliability goals, then changing it.

Technology – Determining that the systems being used in support of the business may be antiquated, then finding and implementing new ones.

Communication – Understanding why communications within your plant never seems to get the message across to the workforce, then changing the process.

Interrelationships – Learning how the various interrelationships within your business either support of fail to support the new change initiatives, then doing something about the problems.

Rewards – Recognizing that the old reward structure may not serve to promote change, then altering it to one that does.

12. 9 Learning and Culture

As we learn and make changes to improve, there is another aspect of learning going on within the organization's culture. Throughout our discussion, we have talked about changing the culture from one focused on reactive equipment repair to a new culture where equipment failure is not accepted as the norm. As we bring in new managers, change the work processes and structure, implement new computer systems, and work through other initiatives designed to accomplish this cultural shift, the organization is watching and measuring the effort. The existing culture, whether good or bad for the business, is one that has been around for a long time. Many of the managers and others who have had lasting careers have achieved their success by working within the current work culture. As a result, the culture and those within it are going to be very skeptical of any new way of doing things. However, new ways can be learned and new processes adapted if the culture can be convinced it is within their best interest.

Everyone of us has been exposed to change programs throughout our career. Most of these could be described as the "program of the day." Over time or because of managerial changes, these programs have come and gone. If we are to be successful with our change programs, we must first change expectations about the program itself. We accomplish this goal by paying careful attention to what we are doing and how it affects the four elements of culture. Figure 12-8 shows how the four elements actually work as part of a system. Failure to address these elements and their interrelationship can spell disaster for change initiatives. Yet if we address them properly, the organization will learn that we are serious and we can more easily shift the culture.

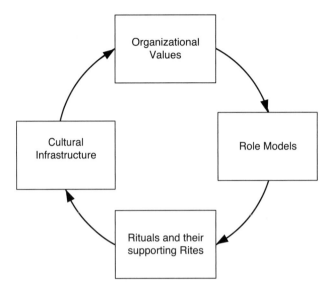

Figure 12-8 Cultural infrastructure model

Organizational Values

The values of the organization are paramount. These values are often expressed as a mission statement, written down, and widely distributed to the organization. If you do only this, you fall far short of the expectations of the culture and what they need to embrace the new way of working. They want to see management "walk the talk." They want to see action in support of the written values, and they want to see it both continuously and consistently. A great deal of effort is needed to establish this level of cultural credibility, but through this effort, the organization learns that you are serious. Examples include:

- Proving that preventive maintenance is truly important by not reassigning the PM crew – ever!

- Learning from failure by trying to understand how to prevent it, not through punishment.

- Telling people training is important and then providing it.

- Praising proactive work efforts far in excess of those who work reactively.

- Hiring people with skills that support the new process.

• Investing properly in better reliability. Cutting costs may look good in the short term, but spells long-term problems while undermining the organization's confidence.

Role Models

Once the value system is in place, the next step in the learning process is to provide the role models who demonstrate the way you want to work. This step can be painful at times as those who had been the role models for the old way of working get moved aside. However, if this transition is not made, the organization will quickly learn that, even though the new values have been established, the new message has no teeth simply because the old role models are still running the show in the old way.

Rite and Rituals

With the values in place along with the role models to support them, the next step in the learning process is to alter the rituals and their supporting rites. These alterations must be made or there will be major disconnects between expected performance and what is actually taking place.

The Cultural Infrastructure

Last but not least is the cultural infrastructure. The infrastructure is anchored in the old way of working. However, if the first three steps are handled properly, those who have primary roles within the cultural infrastructure will learn new things and change over time.

12.10 Learning – A Keystone of The Process of Change

Every organization learns something new every day. The important thing is not just that they learn, but also that they think about what they have learned and then apply it to the process of continuous improvement. If the organization does this, then it will thrive. This applies equally to individuals at every level within the organization. Failure to take advantage of what we learn can be disastrous because we will make the same mistakes over and over. That is why I believe that group (and individual) learning is the keystone of change. Just as the element of leadership drives the process to a vision of the future, group learning is the fuel that gets it there.

Technology

Chapter 13

13.1 Introduction to Technology

When someone hears the term technology in the context of change management and organizational culture, it brings to mind many meanings. In our context, the term is very specific; it refers to the software applications, and the information they provide, in support of the reliability and maintenance work processes in place within your company.

But how does technology, as defined here, involve itself in the study of organizational change at a cultural level? Before answering this question, let's look at some examples. Think about the repair-based method of working. In that process, things broke down and we repaired them, initially using paper work orders and storing the results of our efforts, again on paper. Having detailed information to analyze our problems was not nearly as important as getting the equipment back on line. In this mode of operation, paper was certainly good enough for tracking information.

Now consider a close analogy – unit mechanics. In their most simplified form, unit mechanics are assigned to an operating process simply to fix what breaks down as quickly as possible at the direction of production. Most often these positions are justified in order to keep the small, easy repairs out of the system and to have the simple repairs made quickly. If you look closely, most of these unit mechanics work from a paper system with little or no information retained to indicate what was accomplished. This example shows a low technology / paper system at work.

In today's reliability-focused world of maintenance, our work orders are generated by a computerized maintenance management system (CMMS). The information we obtain to support our efforts is also stored and retrieved when needed from computers. Today we are focused on equipment reliability and the concept that things are not supposed to fail. If they do fail, we don't want only to make a repair, but also to understand why it failed, then make repairs that eliminate the root cause problems so that unplanned failure will not reoccur.

We can now answer the question asked early in this section about technology's involvement with change. Working effectively and efficiently in our new reliability-focused work environment requires timely and accurate information. Meeting this requirement is where technology fits into the picture, both in the systems we employ and the information that they deliver. As used in our discussion, technology is the enabler of change. It supports the other seven elements of change as well as providing a reinforcing foundation for the four elements of culture.

In this chapter, we shall discuss how technology supports change and the elements of change. We will also look at how the proper use of technology can support a reliability-focused culture shift for your organization.

Any discussion of technology must also recognize a "dark side" in the incorrect use of technology and poor information. Either not having or having but not using the technology, then generating incorrect or inaccurate information, will undermine your change processes, also disrupting the new work culture that you are trying to create.

13.2 Technology – The Systems That We Use

Types of Technology Systems

Two types of technology systems support change in a reliability-based work culture. These are administrative software tools and specific functional tools.

Administrative Tools

These are software applications that handle information about the work process. They also provide the user with the ability to work with the information as the work process progresses. Examples include your computerized maintenance management system, a system for controlling supplemental workers (contractors), financial, materials management or financial systems, and programs that provide reports for analysis and action.

Functional Tools

These are software applications that work specifically with a functional part of the business. These tools provide a place to store the information you gather and directly help you conduct problem analysis. They also help you convert the information into data that is analytical and predictive in nature. Examples include vibration analysis, thermography, stationary equipment thickness and remaining life information, operator driven reliability software (usually part of a handheld electronic process), and others.

Reasons for Using Technology

Administrative and functional tools are equally important if we want to change from a repair-based work environment to one focused on reliability. However, many people still believe that state-of-the-industry technology is not necessary. Instead, they believe that doing things the old way, manually or with antiquated software, is good enough. We will discuss this concern further in the section on role models. For now, I want to explain why the old way is simply insufficient if a company wants to survive in today's marketplace.

These older processes worked when all we wanted to do was to fix what had broken down and return it to service as quickly as possible. They do not work in a reliability-focused environment because they can not deliver the essential and required information. What we require are modern systems with information that is accurate, timely, controlled, widely accessible, consistently logical, and integrated. Otherwise, a reliability-focused culture is difficult or even impossible to achieve. Let's look at the key factors required.

Accurate Information

The information within our systems needs to be accurate. Older systems or manual efforts do not always live up to this requirement. Older systems are difficult to manage and information doesn't always get entered. Both manual systems and older computer systems that require extensive manual intervention are subject to human error.

Timely Information

The information needed to make sound decisions may be available, but older systems and manual processes have extreme difficulty getting it to the right person in a timely fashion. Reliability-based decisions require information to be at people's finger tips when needed

Controlled Information

The information within our systems needs to be available, but it also needs to be controlled. The majority of people who need information really only need to see it; they should not be able to remove, alter, or delete it. Newer databases provide this level of information security. Older systems are much more difficult to control; manual processes are even worse.

Widely Accessible Information

The majority of older systems are functional in nature. As a result, they are only accessible to the function and the members of the function that they support. Many people liked this because it was their system and others needed to come to them for their information. If we want the work to have a reliability focus, everyone involved must have access to the information that will lead them to quality decisions.

Consistently Logical Information

Information has a logic to it, especially if it is being analyzed by the system. Software provides consistently logical results. Older software is limited. Human systems are subject to human error or logical processes being incorrectly applied as different people perform the required tasks.

Integrated Information

Older systems and manual processes were fit into functional silos – those in the function were the only ones who had access to the information. Consequently people had to seek out this information from other departments — a time consuming effort — or make decisions without the information.. If you want people to make quality decisions, then you need systems that allow them to gain access to all of the information required to make those decisions. You need to establish a way to bring all of the information together for access by those who require it. This can be a manual process, but preferably you can work within your company to develop an electronically integrated solution — one that links the systems and the information that these systems contain.

Technology tools that provide these basics are not easy to obtain and can be extremely expensive. Some companies have purchased multi-functional software to try to achieve these requirements. The expense is large and the software often fails to deliver one tool for all needs. The problem is that our administrative and functional needs are just too diverse and complex. Other firms simply employ multiple technological tools. They trust that the users of these tools can obtain what they need by knowing how to navigate more than one system or simply by getting the information from someone who does. Each of these approaches is time consuming and difficult. Finally, using new technology solutions, many companies have worked to integrate their systems so that information can be obtained from a single source without having to know the systems from where it came – systems / information integration.

The best use of the technology is obtained by the integration solution. It certainly is less expensive, it can provide a way to link all systems (and the information that they contain), and it delivers what is needed to run a reliability-focused work process.

13.3 Technology – The Information That We Create

Technology in the form of computer systems is only half of the story. When we need to make decisions about how we are going to repair a failed pump, we need more than software. We also need the information that the software can deliver. This information includes:

- pump basics
- bill of material
- what was done the last time the pump failed
- any upgrades planned for this equipment
- the repair history
- safety requirements that need to be included in the work plan

All this information must be accurate if we are to perform our jobs properly. It's the job of the software to make this information accessible and integrated. But that goal will fail unless we have processes in place to make certain that the information entered into our databases is accurate and timely.

Important Rules for Information

Several rules apply to this type of information. If these are followed, the resulting information that becomes available for the decision-making process is likely to be what is required. Without these rules, the information will be suspect by the organization, which will foster a culture that will always be checking its accuracy with the experts. These checks are time consuming and inefficient, and add a layer of skepticism that undermines success. Therefore, you need to apply these rules.

1. Never have the same information in more than one system.

When you have various functions entering information into functionally-specific software databases, you always run the risk of having the same data fields in two places, often with two different values. This conflict needs to be avoided at all costs. First, you run the risk of someone using the incorrect information to make a critical decision. Second, people will learn to distrust the information. The former is dangerous; the latter breeds ineffectiveness and inefficiency. The solution is to analyze the information in each of the databases and remove those that are redundant. There should be only one data set. If the information is required in other databases, create interfaces to get it there on a regular basis from the primary source.

2. Make information accessible in a read-only mode and an easy to obtain format. (Integrate, if possible.)

People need information, but they seldom need the ability to alter it. Therefore, make the majority of the information available in a read-only mode. Most systems today allow this level of security to be placed on the information.

3. Tightly control who can add, change, or delete information.

Engineers maintain the philosophy that people who find inaccuracies in information should be the ones to change it. This line of thinking believes that we are professionals who should have system access to make changes. People probably would not make destructive changes to the information, but it is still possible. More likely, those with good intentions may make incorrect changes.

The problem is that the people we want empowered with the ability to change information do so infrequently. When they do, they are prone to error or to entering the information incorrectly. In addition, the information may need to go into one or more systems; only a data integrity expert familiar with the databases would know the other places the information is needed. In sort, restrict who can make additions to or change your database.

4. Have a detailed process in place for new information, information changes, and information removal.

Along with rule #3 is the need to have detailed processes in place for information handling. If we are going to severely limit who can make additions or changes to the databases, we need to provide a detailed process to enable people to get their updates to those handling the entries. We also need to provide training for entering the information correctly.

5. *Make certain that the information handling process is followed.*
Limiting access and having a process to update information is worthless if the process is not followed. You need to assign a specific person or group from your organization the responsibility and accountability, not only to make changes, but also to make certain that the process is followed. This can be accomplished by contacting the managers of the projects affecting the information and auditing it on an on-going basis.

6. *Evaluate new approaches to information handling and access.*
Finally you need to continually evaluate better ways of handling and accessing the information required for reliability-based decision making. Years ago I was involved with the installation of a new maintenance management system. This system had the capability of storing asset-related details such as pressure settings, weights, metallurgy, and other significant design information. In an effort to make this a value-added tool for the user community, we spent considerable time and money gathering and inputting this asset-related information into the system. Ten years ago that was the best available solution.

Today these fields are no longer utilized for the asset-specific information. Our new approach is to scan the data sheets into an electronic document management system and index the system based on the asset numbering used by the plant. In this way, users who need information can easily obtain the entire data sheet instead of searching for a specific piece of information buried deep within the maintenance software. The ability to update this information through the control of the most recent version of the data sheet is also achieved so that the user knows that the information is accurate.

13.4 Technology in Support of the Elements of Change

As I stated in my prior book Successfully Managing Change in Organizations: A Users Guide, technology is an enabler of the change process. Whether it is through the systems we employ or how we manage information for the user community, technology does nothing more than enable the change process to proceed. Conversely, lack of technology disables change early in the process. In this section, we will examine how technology impacts the change process by discussing its relation to the remaining elements of change.

Leadership

As we have learned, leadership is the key element of successful change efforts. However, for those in leadership positions to promote change in their respective organizations, they need technology in the form of software applications to assure the workforce is given the necessary tools to accomplish the task. They also need to be certain that the information to support these decisions is accurate and timely.

Suppose that as a leader of your organization you want to alter how your firm conducts maintenance. You want to move from a repair-based strategy to one focused on preventive and predictive maintenance. To do this without the proper technology and related information would be a monumental task. As a result, you make sure that not only do you have your work processes in order before you start, but also that systems are in place to support or enable these processes.

Work Process

Work process and technology go hand-in-hand. However, installing technology solutions and then trying to force the process to work with these solutions is placing the cart before the horse. To achieve successful change, you must have your work processes in place first. It is then far easier to find technology that supports the process. Installing technology first limits your options because you are trying to fit a process into a pre-existing frame work.

Another aspect of this discussion is equally important. Software is an enabler of the change process. It is not change. To be successful, you need a process in place that has buy-in from the organization. This holds true even if the process you put in place is manual or uses the current (and possibly flawed) software. In either case, once the organization has embraced the process in its own right, you can more easily bring in the technology that helps the process function.

The same idea applies to the information stored within the systems. First, the process must be in place to capture it and, second, the organization must believe in its validity. With these two conditions met, the organization will be much more open to using the information that the technology presents. Otherwise, it may dismiss the information as inaccurate, then return to the prior manual process for obtaining the information.

Structure

In the past, we often built our organizations around the function that it was to serve. Departments were located near those with whom they interfaced on a daily basis. Members of work groups or functions were all housed in the same building. Other methods were used so that the rule of "form follows function" could be maintained.

With the proper use of technology, these restrictions can be relaxed. Groups can have different structures simply because the work processes are automated by business systems; information is widely accessible. You no longer need to be near those whose information you need to access.

Group Learning

Groups simply can not learn without information. The technology used within your company has the potential to provide this information as well as make it accessible to the group learning process. Without the necessary information, groups will flounder during their learning process. In most instances group learning is about doing something, analyzing the results, and then developing alternate and improved processes and action plans for the future. Without technology to provide the groups with this timely information, they will most likely be unsuccessful. For example:

> How can you plan the execution of a maintenance job better for the future if you can't access the planning and resulting execution information from past jobs?

> How can you accurately predict failure if you don't have a process and system in place to analyze information collected from past failures?

> How can you properly and safely execute a job if you don't have access to the information in the form of design specifications, hazards, etc.?

> How can you predict future problems if you don't have the tools to handle the complex analysis required? Examples of this type of technology include machinery vibration analysis, infrared

thermography, pipeline thickness monitoring and calculation of remaining life, and many others.

Communication

Technology also helps with the communication of critical work information. Having this information at our finger tips improves our performance and the effectiveness and efficiency of the work. E-mail is a prime example of how technology supports change through communication. If not overdone, this system allows people to exchange information instantly and enables improved decision making across the entire operation.

Technology in the form of information also supports communication. Consider how information related to a maintenance job was handled in the past. You recognized that you needed information in order to plan a job. Next you needed to figure out where it could be obtained. You then contacted the people who had it in their office, went to see them, found the information, made a copy, and then returned to your planning effort. Not a pleasant picture. This type of effort was often required several times with several different people in the course of a job. In today's world of technology, this communication of critical job information is far easier. If we handle our information and the integrity of the information correctly, we can gather information and answer our questions almost instantly.

Interrelationships

Interrelationships are also supported by technology. The applications that we use are, in many cases, multi-functional. As a result, groups of diverse individuals from different organizations must work together to provide technology – either new applications or upgrades to those that are already in place. In this way, we are able to leave our rigidly-defined functional department boundaries and work with others to deliver a product for the benefit of everyone.

Because technology is always being improved, your company can reasonably expect that any technological applications it installs will eventually need either upgrades or replacement. Software vendors regularly upgrade their products; it often pays to take advantage of these upgrades. Some vendors add additional functionality whereas others address problems with the existing application. In either case, your organization should pay for vendor support – usually a small yearly fee—and take advantage of these upgrades.

When the technology in use is multi-functional, you often need all of the affected functions to agree to any upgrades. This process can be handled on an as-needed basis, but I would like to recommend an alternate approach. Interrelationships are built over time. Similarly, software applications become an integral part of a business over this same extended time period. Therefore, an internal users group should be assembled to address the optimum use of the technology and develop the interrelationships needed to move the technology use forward. This group should have representatives from all sites (one spokesperson per site). All of the affected corporate functional areas such as materials management, accounting, and reliability should also be represented. Meeting on a frequent basis, this group can support optimal use of the tool and build relationships so that they can collectively position the company to stay current with the software.

Rewards

When one thinks of rewards, one usually thinks of a promotion, a pay raise, or even a bonus. The rewards that technology delivers are of a different sort: 1) software applications that support our work and 2) accurate information that, in the end, enables us to deliver quality reliability-focused decisions for the business. In the world of maintenance and reliability, these things are truly rewards, especially when you think about how work was accomplished in the past and the problems that arose from not having timely access to accurate information.

13.5 Technology and the Four Elements of Culture

At this point we have discussed how the element of technology fits in with the other seven elements of change. At broad view, everything would appear to be fine. You could assume that with the installation of the right technology, along with the entry and management of the information needed by maintenance and reliability, will provide a great deal of value for the user community. However, that is not always the case. All four elements of culture play a critical role in whether the broad view is reality when we get closer to the actual action. For this to occur, technology must enable and support the culture.

Organizational Values

New technology provides direct support for an organization's value system. It contains the information that those working within the system require to make the correct decisions, based on the values. In the world of reactive maintenance, when confronted with a repair, the value system is clear – fix the problem, fix it quickly, and get the equipment back on line. In this scenario, technology does not often play a critical role.

However, let us look at the proactive reliability-focused model. In this case, when faced with a repair decision, the value system requires a much different set of decisions, many of which require timely and accurate repair information. This information comes from the technological systems in place. The value system requires that decisions be made with this information.

Rites and Rituals

As we learned in Chapter 6, rituals are rules that guide our day-to-day behaviors in our jobs and rites are the events or ceremonies that reinforce the rituals. Technology has the ability to drastically change both the rituals and the rites.

For example, consider the implementation of a new computerized maintenance management system – the ultimate technological change for a maintenance organization. When this type of change takes place, the rituals that were in place to support the old process are replaced by new ones that support and are enhanced by the technology. This replacement has a serious impact on the organization and the culture, at both the technology level and within the maintenance process.

On the technology front, the software and the work processes (rituals) associated with it are new. Change brings a level of discomfort and, in turn, resistance. For these reasons, user involvement with the development of new rituals is always part of a successful change of technology. In addition, users are almost always involved in the selection process so that they have a level of buy-in with the final outcome. Further process and software training are always integrated with the effort and rollout of the new system. In these ways, the people whose rituals are being changed are connected to the development of the new set of rituals; they are given the opportunity to adapt to the new rituals by being involved during the entire process.

Because rites are directly reinforcing the rituals, they need to change as well with the installation of a new technology. For example:

A new maintenance system will most likely use improved planning and scheduling techniques. The former process of daily meetings with production to find out what work they wanted done during the day is no longer required. The same is true for the pat-on-the-back rite used for reactively responding to the crisis of the day.

The new system handles bills for equipment and material and also provides the needed parts. It will replace the need to seek out the storekeeper for the things we need to do our jobs (rituals) as well as the praise provided to the storekeeper (rite) for helping us get what we need.

A system provides reliability information about a piece of equipment on which we are working. The system replaces the need to ask reliability engineers to take time from their busy schedules to get it (ritual). It also replaces the process of requesting the information and the time it previously took to get it (rite).

The rituals are changed by virtue of the new technology being put in place. The rites are not so easily altered. They affect how we interact at a very personal level. Often, as in the case of the pat-on-the-back for a job well done, they change how we are recognized for the work that we do. Nevertheless, when you implement new technology, is it unsatisfactory simply to be content with the changes to the rituals and not address the rites. If you fail to take them into consideration, the culture will seek to re-establish a status quo at the level of rites and, even possibly, with the rituals themselves. If you think about this, the culture has its own power to undermine even the best implementation of technology.

Often these problems can be overcome at the ritual level by involving those who will be using the new technology. The rites can be handled in the same manner. We need to include "rite modification" as part of the work process redesign effort. This is something that is most often not even considered when we enter into projects designed to upgrade or replace technology. Using our examples above, we need to consider:

Establishing a daily planning meeting to replace the daily meeting with production to find out what broke down over night. At this planning meeting, production can provide recognition for following the plan versus reacting to emergencies.

Train the storekeepers to be systems experts in the area of bills of material. In this way, they will still provide significant value to the organization. However, the rite will be different; the reinforcement they get will be from using and training others in the technology versus knowing where the parts are on the shelf.

The same approach to the rites of the storekeepers can be applied to the reliability engineers. Instead of being the keeper of the information, they become the technology experts, training others in how to acquire the information they seek.

Roles Models

The individuals in your organization who are the role models play a critical part in the use of technology. There are three types:

The Promoters

These individuals believe strongly in the use of technology to enhance the work process and the reliability of their organization or plant. They are the strong supporters of technology solutions and improved processes.

The Inhibitors

These individuals do not fully support the use of technology to drive improved maintenance and reliability. They are the individuals who believe that what we have is good enough – even if it is highly manual in nature and user intensive. They continually put up roadblocks to delay the delay use of tools that, when coupled with improved process, would otherwise drastically improve performance.

The Statue

These individuals believe that change in the form of new technology is not a priority and that the status quo is sufficient. I call them statues because they do not ever want to change. In many ways, they are similar to the inhibitors because they delay progress.

Role models who fit the definition of Inhibitor or Statue are difficult to change. It may be possible to win their support with a convincing business case, but getting them to see the value of the new technology will not be easy.

In those cases where you can get them to support the acquisition of new technology, great care is needed with regard to the culture. Recall the prior discussions about role models: The organization and the culture existing within the organization take their work cues from their role models. If these individuals are promoters, technology will be acquired and its use mandated. However, if the role models fall into the other categories, the technology — although acquired — may never be used.

Lack of use of the technology you acquire causes several problems.

First, the project fails, costing the organization money and time. Second, there is no progress. Third, the damage done as a result of bad implementation usually creates a situation in which the organization is not capable of additional change for a considerable time.

Cultural Infrastructure

Implementation of technology is a double-edged sword. If the technology is implemented correctly and provides what the users need to properly execute their jobs, it is truly an enabler of the change process. However, the technology could be implemented poorly, it may not serve the users' needs, or, worse, it may make their lives more difficult. In this case, the technology becomes fuel for opponents within the infrastructure not only to undermine the technology, but also to undermine the work process that the technology was designed to support.

Therefore, the members of the cultural infrastructure and the various roles that they play within it are important in successfully implementing new technology or technology upgrades.

Story Tellers

These individuals promote the current culture through war stores about how the culture supported the business. New technology often works with the process changes to support the business in far different ways – for example, the switch from reactive maintenance to planning and scheduling of the work. Story tellers need new stories that focus on the new technology and how it vastly improves the status quo. Because they will not seek these stories on their own, you must provide them. In efforts such as technology rollouts, you must rely on frequent communication showing how the new technology is an improvement. But be careful! New stories must be based on fact; otherwise, the story tellers will see through the ruse and you will be worse off than before you started.

Keepers of the Faith

Because keepers of the faith have the specific task of maintaining the old way, you need to help them recognize the improvements of the new. One way is to include them as a part of the project team. They can then clearly see the value of the change as well as being able to represent that value to the rest of the organization. Because these individuals are viewed as mentors, their support should go a long way to promoting the change.

Whisperers and Gossips

Bad planning and execution of a technology effort, or any change effort for that matter, is food for the whisperers and gossips. If mishandled, they will work behind the scenes to undermine the efforts of the project team. However, if the effort is well planned and well executed, the members of these two groups will have little to whisper or gossip about.

Spies

Spies disrupt technology efforts by taking information out of context, passing it along to their contacts in power, and causing problems as the information is misinterpreted and acted upon. Examples include telling managers that the project is behind schedule without telling them that a recovery plan has been implemented or telling someone the consultant has not developed the training when, in reality, it has already been developed by another firm.

In each of these cases, the work is slowed down. The project team has to stop and conduct damage control when, in fact, there was no damage in the first place. Each interruption is a problem and is further exacerbated when the false or partial information is spread through the organization by the gossips and whisperers. Upfront and complete communication, including the status of the work plan, can often solve this problem

13.6 Conclusion

Technology is an important component of the change process. It enables change and many of the other elements of the change process. However, as you have seen, careful attention is needed in addressing how it impacts the other change elements as well as how it impacts the organization's culture. If handled correctly, technology will provide great value. What hasn't been considered in the past, and must be if we are to be successful in the future, is the way that technology affects the culture.

Communication

14.1 How Does Communication Affect Change?

Communication is an essential part of getting things accomplished in the workplace. Rarely can human beings accomplish anything without the help or support from others. The way we explain what needs to be done is by communicating with those individuals. In turn, they receive and act on our message based on our communication skills. When completed, they communicate this fact to us.

This sounds simple, yet miscommunication takes place every day, both in and out of our business environment. For example,

A mechanic is directed to clean a strainer on a pump that is not shut down.

A production worker sets up a batch process incorrectly because the numbers on the production sheet were hard to read.

An employee returns from vacation and, not being informed about a critical process change, thereby makes a mistake and causes production loss.

A foreman does not tell his crew about a pump upgrade that the reliability group has planned. As a result, they do not include it in the work scope of the repair.

A manager poorly communicates a new procedure aimed at improving work in the plant. As a result, everyone is angered over the change.

A work group explains a simple need to engineering. Not only is the result not what they wanted, but it is very expensive and not usable.

This list can go on and on. One thing that all such lists have in common is that the problems they describe are all the result of poor communication.

Communication plays a crucial role in our day-to-day activities, but it is even more important in the world of change. In any company, change is a major and very significant event. It carries with it a lot of organizational and personal disruption to the status quo; it must be handled correctly. Often with change initiatives there is only one chance to do it right. If you falter during the beginning of a change process, it is often impossible to retrace your steps and make the needed corrections. Furthermore, even if you are able to undo the problem, you often leave the organization very skeptical, making your work and that of your change team even more difficult than it was in the first place.

At the heart of all successful change efforts is good sound communication on many levels. Not only do you want the organization to understand the "why" of the change, but you also need them to understand and implement the various and often complicated parts of the change. Furthermore, you need an open line of communication from them so that you can answer questions and adjust the process to avoid the major pitfalls along the way.

This is further complicated when you consider that change is not always as simple as altering a work process. Quite often there are significant cultural issues that come into play. Again communication is a key enabler of successfully altering a culture. Remember, cultural change involves not only changing values, rituals, and their associated rites, but also having a positive impact on the cultural infrastructure. As a role model for this type of change, you must master the element of communication if you are to be successful. Even if you are not the role model, but are on the receiving end, you still need communication skills to make the change successful.

14.2 The Communication Diagram

Figure 14-1 illustrates the many facets of communication. They must all work together to ensure good communication, whether you are sending or receiving information. Essentially any communication has six components: developing the thought that you want to communicate; converting that thought into a message; sending the message through various media; translating the message by the receiver; receiving the message; and feedback. Each has its own set of pitfalls, places where communication can get fouled up and the wrong result emerge.

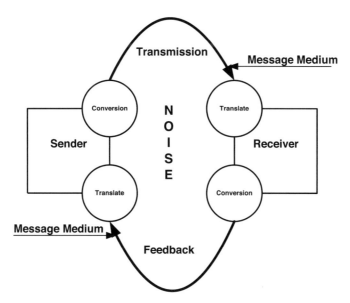

Figure 14-1 Communication model

The Sender

The sender is any person or group who needs to communicate with someone else. The communication could be complex strategic information or an assignment of a simple task. You send information a thousand times a day. This is certainly true in business, especially when you are trying to make a change within the organization. Thus, communication begins when, as the sender, you have a thought that you wish to communicate.

Conversion

You must now take your thought and convert it into a message that you can send to the person or group you want to receive it. Converting a thought into a clear and concise message is often difficult. Think about any time you had to organize your thoughts for a presentation to a group. The process was probably difficult. Even after it seemed you had completed the process successfully, you found that some people in the audience still didn't understand what you meant. The key to successfully converting thoughts into a message is to take the necessary time to make the message as clear as possible. The better that you prepare your message, the more likely is that it will be properly received and nothing will be lost in the translation.

Transmission

Once your thought has been converted into a message, you need to send it. This may sound easy, but it isn't. There are many ways to send the message. How you choose to send it can be as important as the message itself. Media run the whole gamut including oral transmission—speaking your message face to face in a one-on-one setting, speaking in a group setting, talking over a telephone—to written transmission—personal letters, group memorandum, and computer e-mail. Transmission can be in real-time, such as a telephone or instant messages—or delayed time, such as voice mail or letters. If you select the correct medium, you have a higher chance that your communication will be successful.

Translation

When the message reaches the receiver, potential problems can still exist. The message at this point needs to be translated into something that the receiver can and will understand. Unfortunately, the message isn't always translated into the actual information that you wanted to convey. The receiver's frame of reference can cause the mistranslation. Certain phrases may mean one thing to the sender, something else to the receiver. Spoken words can be emphasized in a different way than written words. The feedback stage provides an important way of determining that the translation was clear and accurate.

Receipt

The last part of the trip for the initial communication is its final receipt. This is the step when the receiver processes the message into action. If everything has gone according to plan, then the receiver has received what was sent and acts accordingly. However, if the process is anything like many communication processes, the message is incorrectly received and inappropriate action is taken.

Feedback

The receiver has a responsibility to assure that what was sent was received. This check is done through the process of feedback. Simply restating the message to the sender usually provides this check. The sender can then either confirm that the message was received or provide the receiver with clarification.

Noise

The last component of the communication process is not a direct part of the sequence. Instead, it exists around the process itself and is referred to as noise. Noise refers to everything that is external to the actual communication process, but interrupts or disrupts it. Noise can be simply sound, whether static during a phone call or people talking during a presentation. It can also include office gossip that disrupts the message that people receive or routine interruptions as you do your work. All of these make communication more difficult.

14.3 Important Things to Remember from the Model

When you think about communicating, you probably think either about telling someone to do something or about holding a meeting or group discussion. Yet these uses only scratches the surface of true communication.

Three aspects of communication are important. Depending on the type of communication, each plays a different role.

Sender and Conversion

The content of the communication can run the whole gamut from simple ("install that part in pump P-1") to very complex ("redesign the planning and scheduling process to include more preventive maintenance activities"). Both of these, along with each type of communication in between, bring a different level of complexity to the communication process.

When communicating simple content, a single task is usually required by the sender. The receiver does not have to spend much time interpreting what needs to be accomplished. As the content becomes more complex, the level of difficulty increases throughout the entire communication process. As a result, much feedback and discussion are needed to assure not only that the content was well understood, but also that the actions taken support it.

Transmission

Two basic approaches are available to deliver a verbal communication – oral and written. These two approaches can be conducted in many ways. For example, oral communication can be one-on-one, a group presentation such as a training class or meeting, or a phone call. Similarly, written communications include a personal letter, a departmental memo, and an e-mail. Recognize that the delivery must be the correct one to successfully deliver the content. As the content becomes more complex, the oral or written communication itself can cause a problem because the receiver can interpret the information in many different ways.

Receipt

The last aspect of communication is how it is interpreted – not what you do with it, because that was handled by seeking feedback to clarify content, but how you reacted to it. This aspect is very important because people's reaction to communication, especially about change, can have a major impact on the success or failure of the initiative. For example:

> How do you react to people telling you something if you perceive that they are doing it in a hostile or accusing manner?

> How do you react to a memo about change when you believe that it doesn't make sense or makes things even more difficult?

> How do you react to an e-mail that reads as if the writer is blaming you for something that didn't work correctly?

All of these examples are about how you as the receiver interpret either the content or delivery of a communication. Maybe the accusing demeanor was a result of the person being upset, which had nothing to do with you. Maybe the memo was sent prematurely, before everyone was told why the change was taking place. Maybe the e-mail writer just didn't think about the content before hitting "send".

In each of these cases, no harm was meant by the communication, yet harm was done. If you had the opportunity to seek content feedback, you may have been able to quickly clear up the problem, although maybe not. In either case, behind-the-scenes issues have been created that can pose a significant barrier when expanded over an entire organization trying to go through a change process. As we shall see later in this chapter, receipt of communication also has a significant impact on the cultural infrastructure.

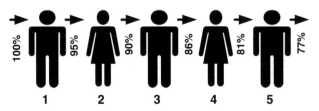

Figure 14-2 Whispering Down the Lane

14.4 The Compounding Problems of Miscommunication

When we were young, we often played a party game called "whispering down the lane." Everyone sat in a line and the leader whispered a message in the ear of the first person. That person repeated the message to the second one in line and so on. The last person in line then repeated the information to the group who, in turn, were amused by the fact that the original message had virtually no resemblance to what was delivered at the end. Why does this happen?

Assume that each player delivers only 95 percent of the message's content given to them by the preceding player. As shown in Figure 14-2, the first person in line gets 100 percent of the message, but it goes downhill from there. The second person receives only 95 percent; by the time the fifth person repeats the message to the group, the content is only 77 percent accurate.

We play this game every day at work in our communication processes. It is bad enough when the wrong equipment part is installed, the wrong amount of additive is injected into the process, or the wrong item is delivered to the customer. However, it is even more devastating when this error happens in a change initiative because the results can cause far greater problems for the organization. The reason is that change initiatives are not communicated in a simple chain as shown in Figure 14-2, but in a more complex pattern of multiple chains. This complexity makes miscommunication even more of a concern, one that requires even more care when we consider what we communicate and how to our organizations.

Let's apply our "whispering down the lane" example to the context of a change initiative gone bad. The maintenance organization of Company X has decided to move from work crews assigned to specific areas within the plant to a single crew centrally dispatched to work on planned and scheduled jobs throughout the entire site. This change has been decided by the maintenance manager in response to reliability problems that have increased throughout the production department. The manager believed this change would allow the correct resources to be applied to the most important jobs.

As before, let us assume that communications effectiveness is 95 percent. Therefore, as it passes through the steps of the communication chain, the new process will be understood by the receiver at only a 95 percent level. Figure 14-3 shows how this communication will be transmitted through the organization and the resulting level of effectiveness

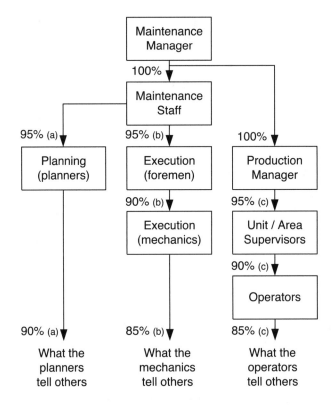

a, b, c each depict the original message but as it progressed through the different chains it ceased to be the same.

Figure 14-3 A communication example

based on the method of delivery.

As you can see, the maintenance manager delivers the message regarding the change in the organization. This message is communicated to the maintenance staff at 100 percent. However, the staff passes it along to the field organization and the maintenance planners at only 95 effectiveness. The planners then relate the message to others at 90 percent. The foremen also pass it along to the mechanics at 90 percent, but in turn they pass the message along at an 85 percent effectiveness level. A similar path takes place in production so that by the time the message reaches the operators it is at 90 percent effectiveness and they pass it along at 85 percent. To complicate matters, the message that has been transmitted is not the same because it has traveled down three different pathways. Consider what you would hear at the end of the three pathways if you compared them to the message that the maintenance manager sent. Then think about how confused a change initiative can become through miscommunication. If you understand the potential problems, you can communicate in a way that assures these problems are avoided.

14.5 How to Increase Effectiveness

The effectiveness of our communication can be improved in many ways. Most take longer than simply telling the staff, who in turn communicate down their respective chains. However, these ways do not allow miscommunication to destroy or otherwise derail a sound change initiative.

To achieve the goal of consistent communication, you need to deliver the same message to everyone. Furthermore, you need to provide a feedback mechanism so that questions can be raised and clarified. The goal is to have everyone at the end of the communication leave with the same understanding of what needs to be changed. Only in that way will change happen as planned.

The Communication Model

Effective communication needs to address each phase of the communication model shown in Figure 14-1. There are many creative ways to accomplish this goal.

Sender

As the sender of the communication, you must be very clear about what you want to communicate. Creating communications in a vacuum is generally self defeating. No doubt you worked in a team to develop the change. With that being the case, you need to develop the communication content with the same team. They (or others) can be the sounding board, bringing in communication-related ideas that you may not have considered.

Conversion

Once you are clear about what you want to communication, you need to convert it into something everyone will understand. This step is often difficult because those working on a change initiative are very close to the action. People this close to the change may understood concepts may not be clear to the masses. You need to be very thorough so that your content is properly converted.

Transmission

How the communication is delivered is critical. The more complex that communication is, the less it needs to be verbalized, and the more it needs to be delivered via a written document with an oral review and feedback session. Figure 14-4 describes how the oral and written parts of communication are balanced based on the type of communication – from simple to complex content. The x-axis in the diagram is content, the y-axis on the left represents the percent that is oral and the y-axis on the right represents the percent that is written. Notice that the axis for oral communication runs from 0 percent to 100 percent, whereas the axis for written communication is the opposite.

A lot of simple content, such as mechanics telling their co-workers "go to the warehouse and pick up the pump seal," is transmitted orally. The written component may be an order form or a piece of paper with the seal stock number. However, in the case of complex content, such as the installation of a new piece of equipment or a detailed preventive maintenance process, the content will be largely written. The oral com-

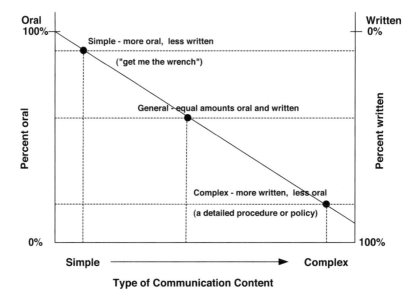

Figure 14-4 Oral vs. written communication

ponent will be the discussion about how to do the work or an actual detailed review of the procedure.

Some communication falls in the middle. These are communications that we wish to convey to the workforce. In these cases, unlike the example of Figure 14-3, we want everyone to leave the discussion with the same information transmitted. In this case, the delivery should be both oral and written. An oral presentation can provide the overview (Powerpoint works well in these circumstances) and a written document can fill in the details. The written document can then be referred to later when some of the meeting content has been forgotten.

Translation

This phase is critical because the communication is now translated into understanding and action by the receiver. Each of us has the potential for translating the communication differently and, in turn, acting differently. Therefore, those communicating must make sure that what they transmit is understood properly to promote action.

Unless an opportunity is provided for those on the receiving end to ask questions, there is no way to assure that the communication is translated as intended. Failure to assure proper communication can lead to disaster in the world of change. Suppose a foreman tells his crew that they are going to increase the level of sophisticated vibration monitoring on the rotating equipment in their area. Possible translations of this could be:

- I will have an opportunity to learn a new skill.

- I will have my job taken by others with this new skill.

- Why do we need to waste our time on this? Our equipment runs fine!

- This is a great idea because it will improve reliability of the plant.

- So what? These new ideas have been tried before and always failed.

- As you can see, translation can lead to many false conclusions about the intent of a change initiative.

Feedback

Feedback is very important to assure that complete and accurate communication is made. Questions, the simplest form of feedback, allow the receivers to restate what they have heard and clarify the areas they do not understand. Questions also allow the communication process to take place again so that those delivering the message can strive to improve clarity and understanding. In our previous example, if the foreman had asked for questions, he would have been able to explain that the new vibration program was being made available to all of the mechanics on the team to improve their diagnostic skills.

14.6 Moments of High Influence (MoHi)

Communications influence, but do not control people. At best we control only ourselves and our actions. We have no real control over others – people make their own decisions and, in turn, live with the result. We can tell workers to be at work by 7:00 or they will lose their job. On the surface this looks like we control their actions, but this is not the case. We influence their actions by telling them that if they don't show up they no longer will be employed, but the ultimate control resides solely with the individuals. The important thing to remember is that we can influence behavior, but not actually control it.

Consultant Paul Balmert has written that in everyday life "there are times and places where people are in a high state of readiness to be influenced." This concept is very important in the realm of communication as well as in the area of change management. Balmert refers to it as a moment of high influence or MoHi for short. At these times, people are very open to be influenced and this usually is handled by communication. Examples include:

- Right after a plant incident

- Upon announcement of a major change

- When something goes wrong

- When something goes right

Recognize that at MoHi, people are ready to be influenced; the communication provided them is critical because it will set up their future behavior.

Suppose a plant has been operating unprofitably and is on the verge of closing down. The management assembles a workforce and explains in detail the plant problems so that everyone understands the outcome if serious change is not made immediately. This is a moment of high influence. People will either agree to change or be out of work. There are numerous examples of how companies have used communication and a MoHi to radically change the business from one that was losing money to one that was an industry leader.

Many years ago I encountered a MoHi. The plant I worked for was told that we would be shut down or sold over the next six months. I brought the entire workforce together and explained the situation. I explained further that the preferred outcome was a sale so that we all would be employed. This was the MoHi. In my communication, I explained to the group what needed to be done to make us an attractive firm for a buyer. We needed to plan and execute our work far better that what we had done in the past. This meeting lasted a long time. There were a great many questions and answers attempting to clear up communications. After it was over, I noticed a major change in behavior and work performance. In the end, we were sold and continued to operate.

As we move through the change process, the important thing to recognize is that there are many MoHi events that will be encountered along the way. When you see one of these moments of high influence, take advantage of them, but make sure that your communications get the message across to everyone as intended.

14.7 Communications and its Impact as One of the Eight Elements of Change

In Chapter 11 of *Successfully Managing Change in Organizations: A Users Guide*, I pointed out that communication was a sustainer of the change effort by being the glue that ties the other elements together. This is accomplished through the transmission of information and feedback for clarification, as expressed in the Communication model shown in Figure 14-1. Let us look at how this affects the other seven elements of the change process.

Leadership

As we have discussed, one of the key attributes of leadership is getting things accomplished through others. The way that others understand what it is that we want them to do is via the communication model. Communication is also the mechanism that they use to seek clarification.

Work Process

When a new work process is established, communication plays a key role. As a change team works their way through the details of the new process, each hand-off of information, data, or tasks assignments, as well as feedback of the results, are critical communications and must be managed as such. Failure to manage this communication leads to confusion about what to do before the new process is even put in place. If this isn't corrected, then the level of confusion is further compounded when the process is actually put into practice.

Structure

As change initiatives are developed, they often require structural changes to the organization. To promote a successful change initiative, you need to consider how the work process will enable or disable the existing communication system. Failure to take structural communication into account can make a simple process unworkable.

Group Learning

In Chapter 12, it was clear from the discussion of single- and double-loop learning that without communication through feedback, neither of these learning mechanisms can work. It is the feedback process in each of these models that enable the organization to communicate with itself and learn from its prior actions. Without the single and double loop processes, there is no feedback and hence no learning.

Technology

This element of the change process is all communication-related. Communication, together with having the correct information at the right time in an easily accessible manner, promotes successful reliability-focused decision making. Otherwise, the communication of needed information breaks down and the benefits that could be achieved are lost.

Interrelationships

Without communication, there can be little or no successful interrelationships. Communication is how we interact with our managers, peers, and those who work for us. However, this aspect is not all oral. Interrelationships are often created or destroyed based on delivery and interpretation of what was delivered. For interrelationships to thrive, you need to pay attention not only to what you say, but also how, when, where and why it is said.

Rewards

Part of the change process is based on rewarding the behavior you seek and not rewarding behavior you want to eliminate. Obviously communication – both positive and negative – plays an important role in this effort. It runs the entire gamut from one-on-one corrective communications to those involving large groups.

14.8 The Impact of Communication on Cultural Change

Communication plays a vital role in the process of change and an equally vital role in changing organizational culture. Communication is the mechanism by which we influence others to change – you cannot make a change successful and long lasting without altering the culture at some level.

Organizational Values

In Chapter 4, we defined organizational values as a company's basic, collectively understood, universally applied, and wholly accepted set of beliefs about how to behave. These values are internalized by everyone in the company and are the standard for excepted behavior. The question is: How do you get an entire workforce to understand and act to accomplish the espoused values of a company? The answer is through proper and continuous communication. If the values are not clearly communicated with an opportunity for feedback and clarification, there is no reason to expect that they will be embraced and acted upon by the organization. To further complicate the problem, the values we are discussing are those associated with changing from a prior way of behavior to one that is new and often alien to the organization.

Suppose that a company has been operating in a reactive mode of maintenance. When equipment breaks down, the organization responds quickly to correct the problem. This company's values are clear to those working in the plant: Fix what breaks and keep production happy. Everything that the company does (both in communication and action) promotes these values.

Now a new manager enters the picture and proceeds to orchestrate a change from reactive to proactive maintenance – where equipment does not break down. This set of values is new and not easily embraced by an organization that had not worked in this manner. A critical success factor in this new direction is communication. The manager and those involved with the change effort must clearly communicate what needs to be done differently. They must include a communication feedback loop so that those impacted by the change can ask questions and get clarification. Failure to communicate the new set of values properly leaves room for misunderstanding, misinterpretation, and counterproductive activity. Without proper communication, the culture will strive to maintain the status quo, resulting in failure of the change effort.

Role Models

The role models in our company are an integral part of the communication process. Role models exhibit traits that we can identify with and use in how we conduct our business. Companies that are making a change in how things are done typically bring in people who exhibit the traits and behaviors they want the employees to emulate, These role models are usually new to a specific plant or company. The old role models – those who exhibited the traits of the former process – are usually gone or moved.

Many firms believe that if you want to make a significant change you need to provide new role models. But these individuals are faced with a difficult problem. They have a process to change and are faced with a culture that is comfortable with the status quo. In addition, even though the company has designated them as role models for the change effort, there is no reason to expect that the culture will accept them as such. Their ability to communicate effectively will be one of their most significant change tools.

Whenever we make changes, the first thing we typically do is hold a large meeting or training session at which we communicate the change and the resulting process. Most organizations believe that, once the process is communicated, they are finished and the process will change as if by magic. This is not so! Making these changes effective is one of the primary tasks of a good role model. Change is accomplished through the process of communication and continual reinforcement of the new process. For our new managers, communication provides a way for them to have the organization associate them with the process change. In this way, as the new process becomes successful, the organizational culture will link the new managers with it, establishing them as a new role model.

Suppose you are promoted to a new plant to change the culture from one that is highly reactive to one that is reliability focused. Your predecessor was considered a role model for the reactive culture. Now you are promoting something far different – proactive maintenance. Over time and through a great deal of work, including a continuous stream of communication from you, the plant begins to change and reliability improves. Because you have been the champion and the process has been successful, the organization will look to you are their new role model. However, this recognition can only be achieved through your communication to promote the effort.

Rituals and their Reinforcing Rites

In any change initiative, the rituals and their reinforcing rites will change. Communication plays a vital role in this process. It is essential that we tell people why things are changing, not once, but numerous times until, through the feedback loop of the communication model, understanding is achieved. Remember: if you don't tell people why something is being done, they will make up their own reasons. Most often those reasons will not be what you wanted communicated. The organizational culture is very good at providing its own reasons, especially if they help preserve the status quo.

The Cultural Infrastructure

Communication is also an important tool of those who hold positions within the cultural infrastructure. As we discussed in Chapter 7, individuals can play many roles including Story Tellers, Keepers of the Faith, Whisperers, Gossips, and Spies. The people who play these roles all use communication either to support the former culture or to support the change. Their importance validates the need for the organization's role models to continually communicate the changes and the reasons for change, especially to those who hold these primary positions in the cultural infrastructure. If the communication is successful, they will pass the information along the hidden communication chains of the culture. If they understand the reasoning behind the change efforts, the message they pass will more likely support than hinder the effort.

Interrelationship

15.1 Introduction to Interrelationships

What do these events within our organizations have in common?

A project is initiated that will add significant profit to the bottom line. However, it needs to be accomplished within a very short time frame. To handle this effort, management brings together representatives from the disciplines involved. There is no time for a sophisticated team-building effort, but a well-functioning team is essential to the success of the effort. Fortunately there is no need for team building. These people have worked togeth er before, they respect each other's skills, and they effectively complete the job.

A multi-functional group works together day in and day out in support of the manufacturing process. At times they all appeared to be doing different things. However, at the end of each day, the process has produced the target quota. As with other teams, they have been assigned to work together even though their areas of expertise are different. They get along, respect each other's abilities and skills, and will gladly step in to support or help the group – whatever the problem.

A staff group working within the reliability and maintenance arena always accomplishes their goals. They share the tasks and stress, while supporting one another at all times.

A group of mechanics work together as an effective unit with far superior collective performance than they would have had if they each worked alone. After work they socialize with one another; it is obvious that not only do they work well together, but they like each other as well.

What these groups and work teams have in common beyond the well-defined traits of a work team is that they have solid, positive interrela-tionships with each other. This enables them to share the work. They believe that for one of them to succeed, they must all succeed; further-more, they care enough about each other to work in this manner.

A group can exist as a team and fulfill the definition and requirements of a team, but still not have positive interrelationships. If this is the case, the team will exist, but it will be less than effective and efficient in accomplishing its tasks and achieving its goals. This chapter will address these interrelationships and see how they fit within the four elements of culture.

15.2 Definition – Positive Interrelationships

Breaking a single stick is in most cases a simple task that you can accomplish very quickly. Now bind twenty five of the same sticks into a bundle and try to break them. Most likely you will not be successful. We are like those sticks. As individuals we may be strong, but collectively as a focused team or organization we are unbreakable. What binds us together in these cases are the interrelationships that exist between us. This strength is what positive interrelationships bring to an organization. However, the opposite is also true. Poor interrelationships essentially sever the binding on our bundle of sticks, until each can be broken in two.

Interrelationships, the seventh of the eight elements of change, is defined as:

> Mutual and reciprocal positive feelings about the members of our organization, department, work team, or colleagues. A strong belief that other individuals with whom we interact are trustworthy and hold the goals of the group above their own.

Think about the people you work with on a day-to-day basis — on work teams, various projects, and other initiatives. Many of these people probably fit the definition above. When you think of your working relationships with them, you know that you can collectively accomplish the tasks that the organization assigns.

Other individuals with whom you interact do not exhibit the traits that make for productive work and positive interrelationships. When you come in contact with these people, your guard immediately goes up because you know that they are not trustworthy. Negative interrelationships make successfully completing work projects very difficult and completing change initiatives impossible.

Many years ago when I worked in the line organization, we decided that we needed a matrix structure for our reliability and maintenance process if we were to be successful. To do this, we created an organiza-

	Mechanical	Machinery	Instrument / Electrical
Maintenance Area #1	**X**		
Maintenance Area #2			

Figure 15-1 Matrix example

tional structure that had both strategic and tactical purposes. We had several functional leads – responsible for the success of their particular function (mechanical, machinery, and instrument / electrical). We also had area work leads responsible for the execution of the planned work in the field. This is shown in Figure 15-1.

In this model, a foreman (shown as an "X" in Figure 15-1) working in maintenance area A1 would report to the area lead and also have a reporting relationship to their functional lead. Each foreman and crew had two supervisors – a true matrix organization. This process could never have been successfully implemented without solid interrelationships among the individuals on my staff. However, I was fortunate to have some of the best people in the organization in these key positions. As a result of their positive relationships, we were successful.

On the other hand, I once worked in a maintenance team with a peer who was known to be untrustworthy and clearly had his next promotion, not the good of the business, as his number one priority. Needless to say, our interrelationships were extremely poor. When this happens at the top of the organization, it always permeates down through the subordinate levels. The result was that little progress was made until he departed.

15.3 Interrelationships and Reciprocity

There is an emotional response between people referred to as reciprocity. When someone gives us something or does something for us, it sets in motion the feeling that we need to respond with something in return. This technique is often used by people trying to get us to do something or buy something that, under normal circumstances, we would never consider. Take, for example, the salesperson who came to my home to conduct a free security and alarm system survey. He spent an hour discussing security and my lack of an alarm system without ever offering a service or a product. What he was providing me was a free assessment, but was it really free? The gift of the assessment set up the reciprocity scenario – my feeling that I needed to give something in return. Then at the very end of the discussion, the salesperson switched from the free assessment mode to the explanation of their services which included a free installation (more reciprocity feelings being established) and a low monthly fee. Whether I succumbed is not relevant. What is relevant is that, in sales of this nature, reciprocity is a powerful force motivating you to "pay the person back" for the service that they have given you. There is no doubt that everyone reading this book has been in a similar situation.

However, reciprocity is not always designed to get you to purchase a service or to do something that you would not do under normal circumstances. Reciprocity can be used to strengthen the interrelationships we establish within the organization or the work group. Here are some important questions about reciprocity.

Why Would You Want To Set Up Positive Reciprocity Bonds?

One of the key aspects of good interrelationships is that people need to perceive you as someone who is trustworthy, a person who places the good of the group above personal gain. Assuming that you are that person, those with whom you work will easily recognize this characteristic through continual day-to-day interaction. However, we are continually placed into work groups that are not composed of our day-to-day associates.

Change projects and other initiatives need diverse resources. At the first meeting, you may or may not know your teammates. Collectively no one knows how they are going to work together to get the project completed. Positive interrelationships are needed from the outset so that productive work can get started. One way to address this problem is through the use of reciprocity. When people feel emotionally compelled to serve the group, the foundation for positive interrelationships is established. The work follows.

How Can You Go About Doing It?

Setting up a reciprocity scenario within a team is easy. All you need to do is to break the ice and encourage others to join in. Offer to lead (or support the leader), take notes, make suggestions that support the effort of the group. Your efforts can even run to the mundane such as making sure lunch is provided. By doing this, you serve the group which, in turn, encourages others to do the same. If they don't volunteer, ask them to participate in various roles. Most people won't refuse. For example, in a team meeting, the leader can ask for someone to take notes, someone to write key points on the flip charts, and someone else to keep track of time. Over time, this involvement builds interrelationships at the basic level, enabling them for the more difficult tasks that the work team will address.

What Is n It for You?

Separate from having successful group work initiatives, there are very powerful benefits for you as an individual by building reciprocity relationships.

People who serve the group are viewed as leaders; they have the ability to add significant value to work efforts because they will be continually asked to participate. Furthermore, by being involved in these groups, you learn significantly more about the overall working of the business.

Successful groups are viewed as doers and achievers. They are often called upon to take on the more difficult tasks. This exposure leads to opportunities for promotion.

The people with whom you work build strong interrelationships that are carried over not only to the next work project but also to day-to-day interaction. You are able to build a network of those to whom you can go for help and support.

You learn from your experiences which, in turn, makes you even more valuable for the next assignment.

15.4 Types of Interrelationships

Everyday we interact with those around us. We develop relationships with people in our organization and those in other organizations. The collective set of positive interrelationships enables us to get things accomplished because no one, except in rare cases, can accomplish much

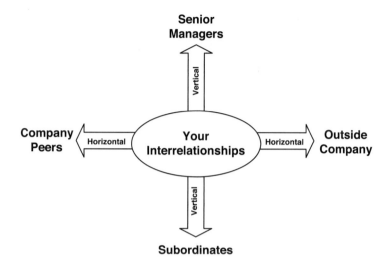

Figure 15-2 Vertical and horizontal interrelationships

in a vacuum. However, each interrelationship is different. Close attention must be paid to each to make certain that, when we need to call them into play, we have built the positive connections. This attention to detail is even more important when dealing with change. There are two types of interrelationships: vertical and horizontal. Within each are specific sub-types. These types are shown in Figure 15-2.

Vertical Interrelationships

Two types of vertical interrelationships are related to the organizational structure in which we reside. These are upward interrelationships with your managers and their peers, and downward interrelationships with your subordinates and those of your co-workers. In either case, when it comes to change initiatives, you need their support.

Manager Approval and Support

From your manager, you need first to get approval for your ideas, and then support to put them into practice. In this context, approval and support are vastly different. Approval is a one-time thing that allows you to move forward. Support is long term. It tells you that your manager will not treat the effort as the program of the day, but will provide long-term support. There are few things worse than a reliability-focused change initiative that has initial support, gets rolling, but then encounters barriers when the longer term support is not there.

Senior Staff Support

Support from the senior staff of which your manager is most likely a peer is a difficult thing to attain over the long haul. One way to accomplish both is to present a compelling business case from which they all will realize value. Then keep them up to date on your progress. If this approach doesn't work, asking for help along the way (even if you may not need it) can help keep their support active.

Subordinates

The same pattern applies to those downward in the vertical relationship hierarchy. Many think that upward interrelationships and downward interrelationships are different. After all, all that you need to do is to tell subordinates what needs to be done and they will do it. Not so! If we have learned anything from our discussion of culture, it is that the cultural infrastructure is extremely powerful; you need the support from all who are its members. Just as with your superiors, subordinates need to buy in to change initiatives. More than those at the top, you need support from those at the bottom over the long run. You may eventually lose your managers' long-term support and still be able to achieve success with change initiatives. However, if you lose support from the bottom, any initiative you are putting into place will flounder and fail.

Horizontal Interrelationships

Two types of horizontal relationships are of concern to you: your peers inside the company and those individuals outside of your company who hold similar positions. This can also include consultants who provide external support as needed.

Peers

The first group of horizontal relationships are peers. These individuals have the same relative position as you within the company, but work in different functional areas. Peers have knowledge in functional areas where you do not. When you consider that nothing in a company can be accomplished in a vacuum, you invariably need your peers if a change effort is to succeed.

For example, take the implementation of a computerized maintenance management system (CMMS). If you are going to implement a new system along with the work process changes that accompany it, then a strong set of horizontal interrelationships will be needed. You will need representatives from systems, materials management, finance, maintenance, reliability, and production deeply involved to achieve a successful outcome. Positive horizontal interrelationships enable all of these functions to pull together into a strong team with positive results.

Those Outside of Your Company

The second set of horizontal interrelationships that are needed are those with organizations, companies, and consultants outside of your company. Just as with peers inside the firm, these people know things and have skills that you don't have and you probably need. Continuing our CMMS example, you will need support from consultants to help with the process change and the software vendor with the system, implementation, and training.

15.5 Trust – Interrelationship Cement

Trust is the cement of interrelationships. It binds groups, teams, and even entire organizations into effective and efficient work groups. Although trust can be established between groups and organizations, it is really the individuals from that organization whom we trust . If the bond is strong, we find it easier to work together because we do not suspect the intentions of those in our work group. We assume that their intentions are honorable, focused on the good of the collective effort. This is a state of group harmony and balance. The exact opposite is true when we are placed in work situations with those whom we do not trust. In these cases, everyone's intentions are suspect. We spend a great deal of time trying to protect ourselves from real or imagined problems.

I have worked with both those I trusted and those I did not. The work interrelationships and the work product were very different. Because we want to have positive work outcomes and successful change initiatives, we must review why we trust some people and not others. We also need to look at how we can build trust with those where it does not exist.

First, we need to make the general assumption that those with whom we interact are trustworthy. This is not always the case as there are people who can not be trusted. In the end. we hope these people will not be those given responsible change tasks. Although this is not always the case, those who are not trustworthy do not make good work group members. They create negativism and seldom deliver long-lasting value-added work. As a result, you are more likely to be on teams with those who are trustworthy individuals.

On the surface, you would think that working with those in whom we have a high level of trust would always be a positive experience, yielding excellent results. This is not always the case, however, because trust is a very fragile bond. It can take a long time to build, but it can be destroyed in an instant and then be difficult, if not impossible, to recreate. One thing that can break trust is when we assume something about someone, then act on it without validation. It may be something that the person said, something that you heard from a second-hand source, or possible even something that was done. In any event, by assuming the worst and then drawing unsupported conclusions from that assumption, the trust bond can be broken.

The corrective action is to not assume anything! Get the facts first before you draw erroneous conclusions. Invariably you will discover that your assumptions were wrong as were the conclusions that broke down your interrelationships.

Years ago I worked with someone who was my peer. We were responsible for the maintenance function in our plant. He ran the execution side of the business and I ran planning and scheduling. We worked well together based on a trusting interrelationship. Then I noticed that he was coming in late and falling asleep at meetings. He was also failing to deliver on work assignments for which we shared responsibility. If I made an assumption about his work performance, as many in our organization did, I would have lost confidence; my trust in his ability would have evaporated. This would have influenced all of our future work and permeated throughout our organizations.

Fortunately I did not follow this route. In order to try to preserve our relationship and our trust bond, I simply asked him about the problem. His work wasn't slipping; he needed help with a family matter. Fortunately the matter got resolved. By not drawing and acting on incorrect assumptions, our relationship was preserved.

15.6 Allies

Recognizing the various type of interrelationships and understanding that trust is the cornerstone, how do we create strong positive connections? We need to build interrelationships, not just for projects, but also for our everyday work. We need people who are allies – those with whom we have personal trusting relationships and who will support our work. These are the people you can count on to help achieve change and, ultimately, cultural shift. There are two types of allies – the ones that are created over time and those that are built over the short term.

As you work in different positions throughout your career, you interact with a great many people. Those with shared values and with whom you have built positive interrelationships based on trust and respect become allies for the future. Some may be in the background, available to come to your support if needed. This relationship is reciprocal; it also applies to you supporting them if they are in need as well. Although you can't expect them to "take a bullet" for you, you can count on them for support and advice. I have had the opportunity of working with many fine people over my career. As a result, I have built strong alliances. Those that have stayed in the company now hold positions of responsibility. Having allies of this sort has paid off in getting things done far more efficiently than otherwise possible.

Although it often takes years to create long-lasting alliances (usually done over the course of our careers), expecting that this longer-term approach is the only solution to building alliances is flawed. Not everyone has the time to work at a slower-paced fashion; many do not stay within a company long enough for alliances to take hold. Therefore, allies must be developed over a much shorter time period. They are often created based on common need. For instance, two people may be working on a project for which they have joint accountability. Obviously a possible short-term alliance could result. This scenario also applies to change initiatives and the associated cultural shift. In these cases, we need the alliances, but don't have the time for long-term interrelationships to develop.

Because shared values, a common goal, and mutual trust are the cornerstone for a good alliance, and because you need to establish allies over a short-time period, I recommend that you take the initiative to create your allies. The common goals are already established as you most likely are working together on a project or a management-directed initiative. Shared values will emerge as you work with others. I am not saying that you need to have exactly the same values. But you will need to share common beliefs about work and how things get accomplished. Two people working together, who both believe in involving multiple layers of the organization, share values about involvement and can build on this value. However, if you believe in involvement and a person with whom you are working believes you can mandate change, a set of shared values may not exist, making an alliance difficult.

Building alliances is much easier than it appears. You can use the following five practices to help you. Although I have recommended that you take the initiative and start things rolling, realize that for alliances to form they must be two-way. If they are not reciprocated by your potential allies, they will never take root.

Act Professionally at All Times.

A person who is judged to be a professional in word and deed is usually someone who is consistent in their work. The last thing someone would want in an ally is someone who is unpredictable and unprofessional in approach.

Keep Your Promises.

Trust is built on saying what you will do and then doing what you say. This reliability promotes consistent behavior, clearly telling those with whom you are allied that what you tell them will be delivered as promised.

Treat People as Equals.

There is no easier way to ruin potential alliances than to violate this rule. Alliances are built on perceived equality. People do not like to be made to feel inferior; they will not want to ally themselves with those who treat others in this manner.

Go the Extra Mile.

Not only should you be counted on to complete the work you promised, but people with whom you create alliances should reasonably expect that you will devote nothing less than 100 percent to the effort.

Communicate

This rule is the most important one for building and maintaining solid alliances. Not only should you communicate, but also you should make sure through the communication model's feedback loop that your allies have received the message as you intended it to be received.

15.7 Interrelationship Forces and the Other Elements of Change

Interrelationships are an important element of the eight elements of change. They need to be carefully considered when working with any of the other elements. Suppose you are working in a plant with a very reactive maintenance work philosophy. The plant's best practices are tied to how quickly the maintenance organization can fix equipment that breaks down. There is very little room for proactive maintenance because everyone is focused on the "break it – fix it" process. Then the plant is sold; the new management team does not share the same beliefs as the former owners. They believe that things should never break. Processes need to be in place to identify potential failure and provide time for the organization to make the repair proactively, thereby minimizing production downtime.

For the new management team to make a successful change from reactive to proactive maintenance, interrelationships are critical for each of the elements.

Leadership

As leaders of the organization and agents of change, the new leadership team needs to establish good interrelationships both horizontally and vertically. They need buy-in from the organization (vertical) as well as support from those in other departments (horizontal). Without these interrelationships existing, any change effort is doomed.

Work Process

Most certainly the work process will change. Processes associated with reactive and proactive maintenance are not the same. However, changing the process is not as simple as redrawing the work flow. Work flows are built on interrelationships; for a new process to take hold, they must be addressed.

Structure

Just as the work process will need to change, so will the structure in all likelihood. Once again, interrelationships play an important part in this effort. If the interrelationships are not addressed, along with a new set of roles and responsibilities, the new structure will prove ineffective. A new structure coupled with a new process will drastically alter the old interrelationships.

Group Learning

This element is based on learning as teams. Interrelationships again play an important part. Teams learn as they progress; positive working relationships support this process.

Technology

In today's world, most change is supported by technology – applications that are designed to support the effort. For this element, positive interrelationships need to exist with the various systems organizations that deliver this support.

Communication

As we have seen, communication is closely tied with interrelationships. With good communication, interrelationships are strong and both short-term and long-term alliances can be built. Without good communication, these alliances can not properly be built.

Rewards

As we shall see in the next chapter, rewards are not all about money. Positive working interrelationships deliver a different type of reward – mutual support and successful work initiatives.

15.8 Interrelationships and the Four Elements of Culture

Our goal in Chapters 9 through 16 has been to describe in detail each of the eight elements of change as well as tie them into the four elements of culture so that the relationship is clear. Just as interrelationships play a key role in the eight elements of change, they play an equally important role in the four elements of culture.

Values

Shared values drive cultural change. Values can be established by senior management. However, only through good interrelationships and trust will they be accepted, agreed upon, enhanced, and disseminated by the culture within the organization. The easiest way to appreciate this is to consider our prior example of changing a reactive maintenance process into one that is proactive. Suppose that the interrelationships between senior management (the change driver) and the rest of the organization are poor. Suppose too that the new management team runs the business by mandating changes, then punishing those who are perceived not in compliance.

Needless to say the interrelationships in this case will be poor. In this scenario, how successful do you think the management team will be in delivering and having accepted a set of values that is essentially alien to the organization? Without good interrelationships, the new value system will never be adopted. Change may take place on the surface, but the culture will maintain the status quo behind the scenes. There is little doubt that the management team will eventually see this because reliability will not improve. The "break it – fix it" mentality and work practices will remain and their efforts at positive change will fail. A culture thrives on good interrelationships. Without them, change will not take hold.

Role Models

The organization's role models are those who support the value system. In doing so, they will be successful. The message they send to the culture is that rewards are the result of achieving the company's goals and these goals are driven by the organization's values. Over the short term, some role models do not support the value system, especially when it is new, such as switching to a proactive maintenance process. However, once their lack of support is discovered, senior management will remove them from positions that enable them to be role models for the wrong things.

Role models can't be effective without good interrelationships both horizontally and vertically. The relationships and trust that role models have with others is an important part of their function. These positive relationships and the success of the individuals sets them up within the culture as role models.

Continuing with our example, suppose that, in order to promote their change initiative, the new management team places new individuals in the leadership positions within the maintenance and production departments. These individuals are between "a rock and a hard place." They need to deliver on management's new vision of proactive maintenance, but certainly they can never accomplish this task alone. They need the organization and the work culture that supports it to embrace the change. That support can only be accomplished by building positive interrelationships. In building those interrelationships, they promote the goals. Through a successful outcomes, they have the potential to establish themselves as the new role models for the organization.

You might wonder what happens to those who previously occupied the role model position. Many of them will align themselves with the new process and thereby continue to hold their role model status. This transition is very beneficial for the success of the new process because existing role models are easier to follow than new ones who you are trying to create. The culture will view the old role models' switch of approach as necessary; the transition will be easier. Those who can not align themselves with the new process will cease to be organizational role models.

Rituals and Their Supporting Rites

Rituals and their supporting rites are what make up a work process. Positive interrelationships are required for successful completion of any one of the work process steps (the rituals); they are also what binds the various parts of the process together into a successful whole. Consider how well a proactive work process would work if those involved in each of the sub-parts could not get along and did not support the work of the group. In this case, the rituals would break down and the results would be far less than satisfactory. This also applies to the supporting rites. If there are poor interrelationships, the rites that reinforce the rituals may never take place. This lack of reinforcement tends to diminish the effectiveness of the process as well.

Cultural Infrastructure

Four groups within the cultural infrastructure must be addressed in our discussion of interrelationships. Those who play a primary role in the cultural infrastructure all have strong interrelationships throughout the organization. These interrelationships partially explain how they have attained their positions within the informal organizational structure. Because they hold positions of high influence, they can affect the beliefs of others. Their interrelationships can be used in a positive manner to communicate and support the change process. Let us look at how the interrelationships that exist between these people and the balance of the organization can support our efforts.

Story Tellers

By telling their war stories, such as how maintenance stepped in to save the day when a critical piece of operating equipment failed, these individuals are promoting the status quo. We need to create our own stories that significantly overshadow those that promote the old ways of doing business. For instance, communicating stories about reliability and the fact that improved reliability helped the company to make a record profit can help people to recognize the value of this new type of work culture. These stores will then be viewed as more significant than those about failure. Over time, the story tellers will adopt the new ones.

On the surface, this may seem simple, but it is not. Remember the story tellers are working with a great deal of history whereas you are trying to make a change to a new way of working with no history. To accomplish this switch in the stories, you need many new ones that reflect the value of the change. At first, you may even have to become a story teller yourself. The message that there is more value in the new way of working needs to get out to the organization.

Keepers of the Faith

These individuals, who protect the existing culture, educate those new to the organization in the old ways. This contact builds strong inter-relationships that place these individuals in positions of cultural influence. If you want them to become keepers of the new faith, you must involve them in the new process so that there is a feeling of ownership from the outset. The first step in this process is to identify who they are and then get them involved. Because their primary role is to protect the old, this change will be difficult for them. As they come to grip with the new way of working, they may still prove difficult for you because there will be resistance. This resistance can develop from their own fears about losing their position.

Whisperers

Whisperers have power within the infrastructure because they have access to the managers' ear. We can only hope that they whisper the truth, but often they do not. Nevertheless, whisperers are part of an underground communication system. They draw power because they have high-level access and maintain their interrelationships so that their power can continue. If managers are focused on changing the culture, then information from the whisperer community can be of value; it can be used as a barometer for measuring the progress of the change initiative. However, if you are on the receiving end of this information it must be distilled. You need to be able to sort fact from fiction.

Gossips

Gossips make up the hidden day-to-day communication system of the firm. Just as with whisperers, much of what they pass may not be factual, but they will give you a sense of what is really going on within the organization. There are ways to tap into this communication system by building good interrelationships with the people involved. A good manager will walk through the plant and talk to the people. Once the formal process has been made public, people will talk and tell you what is really on their mind. Although this may not happen immediately, it can happen if you listen and demonstrate concern for the issues of others. Knowing the gossips and linking into their communication system through positive interrelationships with them can give you another view into what is really going on in the plant.

15.9 One Last Thought

Organizations are nothing more than groups of people focused on common goals. They share values and work together through positive interrelationships. The interrelationships within the workforce and those between management and the balance of the organization are of critical importance. They are the glue that holds the organization together, allowing us to move forward, make changes, and successfully achieve our vision.

Rewards

16.1 Why Consider Rewards?

When people initially think about rewards in a work context, they often think about money or a job promotion that then leads them to think once again about money. Rewards, however, are far more significant than simply how much money someone gives you for a job well done. As we shall see, monetary rewards are short-lived and do not always add the value that we perceive. The rewards we will discuss address successful organizational change, reliability-focused culture shift, and work process improvement.

But why are rewards one of the eight elements of change? The answer to this is that we can use rewards to reinforce the things that we want to get accomplished. Similarly, we can withhold rewards for those things that we want to eliminate.

For example, take a reactive maintenance work process in a continuous process industry. The first thing that the maintenance group does when they arrive at work is to check what equipment broke down during the night shift, then determine which of these failures threatens safety, environmental conditions, or plant production. Next, the maintenance organization mobilizes itself to rapidly fix those critical things that are broken. At the end of the day, if this process has worked smoothly, the emergencies have been corrected.

As a result of these efforts, what does the organization do? Invariably it praises and rewards those who have reacted to and resolved the problem-of-the-day. This praise carries great value for the planners, foremen, and work crews who have saved-the-day and kept the plant in operation. Although the reward may show up later as a better salary increase at the end of the year, at the moment it is the praise that carries the value.

I know from experience, when I was a field execution superintendent, I was one of the best reactive maintenance supervisors around. Then I recognized the error of my ways. At first, I followed the exact process I have described above. As a result, I received the praise I sought and went home, ready to repeat the same activities the next day. My reward was the praise from our customers. In fact one of my reactive work mentors came into work on weekends so that he could handle maintenance-related emergencies and get praised on Monday. He was not even paid for the weekends he spent in the plant. In our culture, management rewarded the reactive work process and those who performed it well. They reinforced the behavior and "what got rewarded got done."

What do you think would have happened to me if, being presented with that day's set of problems, I told production, "I am sorry we are not working those jobs today. They do not appear to be emergencies. We are sticking to the work scheduled." In that case, management would have applied negative reinforcement to remove me from my position, hiring someone else in my place who fit their expectations of a reactive supervisor. The negative reinforcement would also have been a message to the rest of the organization.

By contrast, consider proactive maintenance. Now we perform work so that equipment does not fail. We conduct preventive and predictive maintenance tasks and are able to identify failure before it happens. The reward for this type of work is more subtle and not as rapid as under reactive maintenance. Over time, equipment fails to breakdown. Those who understand proactive maintenance recognize that the work that has been done to prevent failure has paid off. When equipment reliability improves, the maintenance group receives some praise, but it is not as timely as the praise received by the repair-based culture. At times it may even seem as if high levels of reliability are simply expected as the norm. This expectation can make driving reliability-based work processes difficult. The challenge becomes even greater because when a failure occurs, the praise often goes out to the crew who saves the day rather than the crew who repeatedly prevented failure on other assets, and delayed the failure on the equipment in question.

Thus, rewards are one of the eight elements of change. We need to understand how to reward proactive behavior over repair behavior so that we can successfully drive change.

The four elements of culture further compound this challenge. The values we establish support how we run our business and what we do or do not reward. The role models within our work processes are those who do or do not deliver the rewards and drive the process changes. The rituals and their supporting rites are also an integral part of the work and reward process because the established rituals are what management rewards through the reinforcing rites. Last but not least, the cultural infrastructure applies subtle behind-the-scenes pressure by rewarding compliance to "how things are done around here."

16.2 Short-Term and Long-Term Rewards

The previous section described reward scenarios for both reactive and proactive maintenance. Because rewards are dispensed for work accomplished according to expectations set by your organization or manager, you need to understand the inherent problem of proactive work rewards.

Figure 16-1 describes our repair-based work scenario. First, something breaks down (block 1); then the problem is identified, a maintenance crew dashes in, makes the save and order is restored (block 2). As a result, the crew receives praise from production for a job well done (block 3). The praise is most often immediate and the reactive behavior is, therefore, immediately reinforced. Maintenance is now ready for the next emergency (block 4). This process is made possible because the time delay between the work and the reinforcement is minimal. This

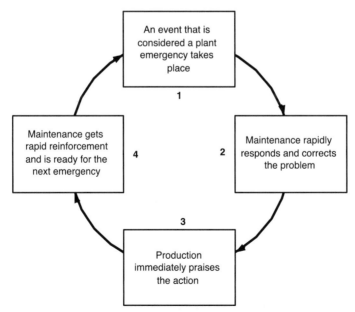

Figure 16-1 Reactive rewards

close coupling of the action and the reward is an optimal way if you wish to reinforce behavior.

Now consider the preventive work scenario shown in Figure 16-2. In this case, the preventive maintenance efforts can go on for years (block 1). The result is no equipment failure (block 2). When this success of the efforts is recognized (block 3) because the amount of planned or emergency repairs has decreased, some praise may be given (block 4). The problem here is that at two places in this cycle there is a long lapse between the work and the praise (if it is given at all). First, there is a time delay between when things actually stop breaking and production recognizes the improvement. Second, there is another time delay between when the improvement is recognized and praise is provided. These time delays make the praise and the reinforcement it is supposed to provide less than effective.

These time delays also make reinforcing proactive maintenance behavior very difficult. This challenge is also true for any change you wish to make which has a disconnect in time between the actual work and the reward.

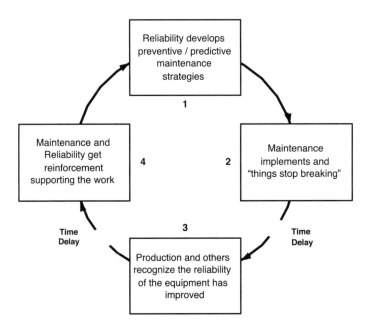

Figure 16-2 Proactive rewards

Think about the performance review process. Most companies hold these events on a yearly basis. At this time, you are called into your manager's office and told how you performed over the past year. Often this review is accompanied with a salary increase. The problem here is the same as that described in Figure 16-2. The time delay between the actual work that you did and the praise for the effort is so far disconnected in time that whatever praise you receive has little or no meaning.

Even for long-term work efforts, short-term rewards and praise are needed. They let people know where they stand in shorter time increments. If employees have done well, they receive the rapid feedback they need to re-energize their work. On the other hand, if they have not done well, they have sufficient time to take corrective action and modify their work behavior.

16.3 Rewards and the Hierarchy of Needs

What types of rewards motivate people to do the things that they do? Although your initial thought may be that money is the main driver, time has proven that this is not the case. This insight is important, especially if you provide salary increases or bonuses hoping to implement successful change, but are surprised that nothing different happens.

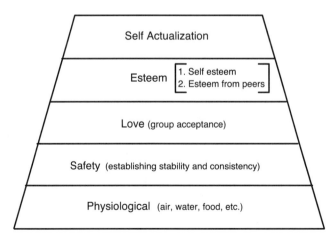

Figure 16-3 Hierarchy of needs

Maslow's Hierarchy of Needs

To understand the real process of human motivation, we need to consider Maslow's Hierarchy of Needs. Abraham Maslow believed that "human beings are motivated by unsatisfied needs, and that certain lower needs need to be satisfied before higher needs can be satisfied." Maslow established five distinct levels of need, as shown in Figure 16-3.

These levels are described (from bottom to top) as:

Physiological
This level covers basic needs such as air, water, food, and sleep.

Safety
At this level, people need to establish stability in a very chaotic world. They meet these needs through home and family.

Love (non romantic)
This level of needs describes our requirement for belonging and being accepted. We fulfill this need by joining groups, clubs, and other organizations.

Esteem
There are two types of esteem needs. The first addresses our need for competency or mastery of a task or skill. This need is for self-esteem. The second addresses our need for recognition of these competencies by others, especially our peers.

Self actualization
This level addresses our need to achieve our potential. Each of us wants to live up to and exceed our own expectations for ourselves.

Satisfying the Hierarchy of Needs

The Hierarchy of Needs theory states that you must first satisfy your lower level needs before you can move up the pyramid. For example:

How much effort would you put into becoming an accepted member of a work group (non-romantic Love) if you were constantly worried about losing your job (Safety)? The answer is very little because you would be preoccupied with satisfying your need for Safety – a lower-level need on the hierarchy.

How much effort do you think you would expend learning to be an excellent planner (Esteem) if you could not get recognized as an integral part of the planning team (non-romantic Love)? Once again, not much because you would first need to become a member of the team before putting out the effort to become an expert.

These are but two examples of the need to satisfy lower level needs before moving on to satisfy needs that are placed higher on the hierarchy.

Applying the Hierarchy of Needs

The Hierarchy of Needs directly relates to how money is used as a motivation tool for work improvement and, more specifically, for change. The majority of us who have a job and a reasonable salary do not expect to lose these safety-related items. Consequently, a salary increase rewarding yearly performance is not by itself a motivator of change. It sits somewhere at the Safety level of the hierarchy. To complicate matters, the salary increases are usually given once per year. Using Figure 16-2 as the model, we can recognize the significant time delay between the actual performance and the salary increase. Because our safety requirements are already met and our work group satisfies those needs of non-romantic love (belonging to a group), salary increases have little motivational effect on change.

For change to be successful, we need to motivate our workforce by providing rewards that are at the levels of non-romantic love and, ultimately, esteem. This goal can be accomplished through work group assignments and work content assignments. The former gets the individuals into a group where they and their skills can be accepted. From there, they can then further develop those skills and achieve the esteem – both externally and internally — that will support the change initiative. Work content assignments give the individuals the ability to learn new things and excel in their performance. As we have learned, this achievement satisfies both self esteem and esteem from one's peers.

Let us assume that we work in a plant that does not have a planning and scheduling function. Figure 16-4 depicts the process that a team would go through in the development and rollout. In block 1, the site determines that a planning and scheduling function is needed. In block 2, selected key people are assigned to the work group responsible for the development of this new process. These individuals are motivated by the self-esteem factor. If their assignment is handled properly by management, they will feel that they have been selected because of their expertise and ability to work on new and innovative projects. In block 3, the

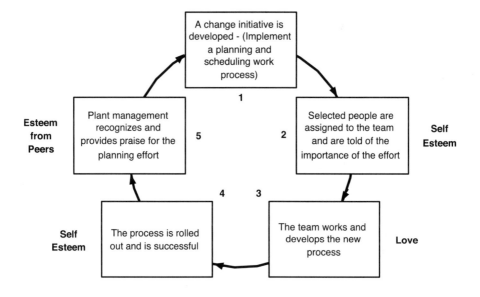

Figure 16-4   An example of how the hierarchy works

team works and develops the new process. As teams of this nature work together, they build bonds that usually last long beyond the end of the work initiative. Furthermore, being an accepted member of an important group working on a key management initiative directly addresses the need for Love (group belonging).

In block 4, the group delivers and rolls out the process, which is successful and immediately provides significant value to the plant. Seeing the success and knowing that they were each a part of it provides a great deal of self esteem to the team members. Finally in block 5, the work effort is recognized by plant management and one's peers, providing the second part of the esteem need shown in the hierarchy.

Another aspect of this model overcomes the problem of time delay seen in our prior example: At each phase of the effort, starting with block 2, a reward is provided – although the reward is different at each level. Therefore, even the time delays that occur between conception of the idea, its development, its rollout, and the recognition by management do not cause a problem for those involved in the process.

16.4 Negative and Neutral Rewards

Not all rewards are positive. Some have a negative effect while others have no effect at all. Because the reward system is built to motivate organizations and people, these types of rewards are not desirable in any context. Every one of us has been involved with rewards to this type. Negative rewards include:

You work long hours on a project, getting home late and having little time to spend with your family. At the end of the project, management decides that the reward for the long hours away from home is to go to a long dinner after work.

You are recognized as an expert in a certain area and management has made no effort to develop any back up. As a result, you are assigned to every new initiative that addresses your area of expertise. Each time you are assigned, you are praised as someone who is critical to the effort's success.

At the end of a very productive year, you receive the standard raise and are told it is the fault of the salary system that you can not receive more.

You are one of the best mechanics in your plant. In a typical year, you work 25 percent overtime. The extra money has really helped you financially, especially because your child is starting college next year. Now because of your expertise, you are assigned to a project that will virtually eliminate overtime.

Neutral rewards include:

You have been with your company for many years and are earning a very good salary. Your yearly raise is 2.5 percent.

The company decides to send the key employees to a training program. The plant population is told it is a reward for select employees to learn a new skill. You are sent, but you have been doing the work covered in the training program for several years and there is nothing new.

The reward for completing your project on time is baseball tickets. You hate baseball.

If you consider the rewards listed above and also consider that the goal is motivation, you can easily see how these rewards fall short of the mark – many of them very short. Motivating people to implement new initiatives and change how things are done requires careful thought before the rewards are determined. For each of the negative and neutral rewards listed, there is a way for those creating the reward to turn something that is negative or neutral into something that is positive. If this isn't the case, then the reward may be entirely wrong for the goal you are trying to accomplish.

16.5 How We Reward

Rewards should motivate people to performance in a way that promotes the success of the business, the work group, and even the work process that the group is using. Larger rewards most often go to those who add more value than those who simply perform their jobs. Negative rewards often go to those who do not measure up and fail to perform as required. We want to model behavior and we plan to reward people for following the model. Therefore, we need a process that makes the performance indicators very clear to those who are being asked to comply. We also need a process that continually provides feedback to people so that they know if they are on track. This feedback is even more important if the measures are addressing change initiatives and simply performing the existing processes well is no longer satisfactory.

Chapter 3's discussion of the Goal Achievement Model suggested the process for establishing clear measures of performance. Management, or those wishing to make the change, set the vision, the goals, and the initiatives. Through a set of discussions with the work team or individuals engaged in the initiative, specific activities are then established along with the measures that ascertain completion of the activities.

The activities are often represented as milestones along a more involved process; they almost always have dates associated with them. The manager and work team or individual can periodically review both the Goal Achievement Model and the measures associated with the activities to determine if satisfactory progress is being made. If not, timely corrective action can be taken. However, if the effort is on track, new activities and related milestones can be established. In addition, this is the opportunity for the manager to provide rewards in the form of recognition.

Referring to Chapter 12 on group learning, you will quickly see that this process is a form of single loop learning. Individuals can determine how they and their team are doing in accomplishing the initiative and its related goal. If conducted correctly, this process clearly communicates the desired outcome of the initiative, directly holding those involved responsible and accountable. Furthermore the process provides timely feedback about the effort's success.

16.6 Rewards and the Other Seven Elements of Change

Rewards are the eighth element of change. As we have learned, it is very important in its own right; however, it plays a key support role for the other seven elements. The thought of reward (or the potential lack thereof) is a constant force when we are involved with any of the other seven elements. Total failure to accomplish the assigned task almost always brings with it the type of negative rewards we never want to receive – demotion, reassignment, or job loss. On the other hand, great successes most often bring with them promotion, recognition, and salary. Consequently, we are always aware of what we are doing and the potential outcome of our effort. Rewards impact each of the elements as follows.

Leadership

Because plant leadership not only sets the goals and initiatives, but also provides rewards for those who accomplish them, rewards are an important part of the leadership's tool kit. The key part of the process that must be well managed, especially if leadership is championing change, is the clear communication of these goals and initiatives. In addition, an on-going process must be in place to help those who are trying to achieve them to understand their progress and also to receive redirection if needed. This continual feedback is a part of the reward process. When my manager tells me that my work team is on track and that we have done very well, that information feeds into several elements of the Hierarchy of Needs – non-romantic love and esteem. However if there is no reinforcement, we will not be clear if we are delivering a work product that meets management's expectation; our ability to deliver a successful change initiative will be impaired.

Work Process

Change initiatives usually include changes to the work process. Because change is such a significant event, the proper application of rewards is important. One area of rewards that requires specific attention is that of time delay. Those leading the change effort must recognize that any excessive time delay between change events and reinforcing rewards must be minimized. As changes are put into place, the work team and even the entire organization are looking for reinforcement. If reinforcement is not provided, assumptions will be made about the level of success. These assumptions most often lead to the wrong conclusions and, ultimately, wrong actions.

Structure

An important component of any change effort is how the new process often requires the structure to be altered. As with changes to the work process, this effort is often significant, with a great deal of stress. Timely reward for small successful changes will allow you to send a clear message to the organization that things are going well.

Group Learning

Rewards are very closely tied to Group Learning. Groups take action, review the outcomes, and learn from what they have accomplished. Rewards, in the form of recognition by management, send the message that the learning has been accomplished. In addition, group belonging is strengthened; both self esteem and peer esteem are built. Think about a successful team effort in which you were involved. When completed, you received praise from your management team – that made you feel good. However, what probably made you feel even better was the mutual recognition from your team and your peers as well as your belief that you had accomplished something significant.

Technology

This element is not significantly impacted by rewards. In a way, technology is similar to rewards in that it provides support to other elements. However, technology has enabled companies to implement on-line measurement processes that provide a significant level of support towards making the measurement and reward consistent. Many companies now have a process like the Goal Achievement Model computerized so that key measures of success can be tracked consistently across large plant and company populations.

Communication

Rewards require communication if they are to have any chance of being successful motivators of change. Not only does management need to clearly communicate the goals and initiatives. but they also need to communicate the level of success that teams and individuals are having achieving them.

Interrelationships

Establishing sound interrelationships across work teams and organizations directly supports the Reward element. It is these interrelationships that support strong group membership (non-romantic love on the hierarchy) as well as peer esteem for a job well done. Without sound interrelationships, both of these levels of the Hierarchy of Needs may never be satisfied.

16.7 Rewards and the Four Elements of Culture

As we have seen, rewards can have a real impact on how well change is implemented and, ultimately, succeeds. Group memberships, selection for change project teams, and praise that builds self- or peer-level esteem are all very powerful forces. Similarly, rewards that are used to help successful change initiatives carry the same level of power into the organizational culture.

Organizational Values

Values are at the very heart of any organization's change process. If an organization values and rewards reactive maintenance, then it is difficult to believe that a proactive initiative such as preventive or predictive maintenance will have much chance of success. In fact, those in the organization working on proactive initiatives will feel punished because they will be out of the mainstream, not receiving the rewards for reactivity being given their peers. However, if an organization and its leadership clearly articulate a new set of values – ones focusing on proactive maintenance – and use rewards to reinforce their implementation, then success is far more attainable.

In order to show clearly how the new values are being rewarded, you can use the Goal Achievement Model. Using this tool, you can provide the vision along with the new value system and goals needed to achieve it. The organization then is able to develop the initiatives and activities needed to reach the goals. This coupling of a new set of values with a clear set of goals via the Goal Achievement Model sets up a measurement system to track and reward successful change. If the Goal

Achievement Model is used correctly, those within the organization will quickly recognize the requirements for success and reward.

Role Models

Two aspects of role models need to be discussed. The first addresses when our leaders are also our role models. In this case, having the people who set the direction for the business also be the ones we wish to emulate provides a tremendous benefit as the leaders work to promote change. As long as they practice what they preach — and what they preach is the strategic direction for the organization — the leaders will be followed. In turn, the organization and its supporting culture will change.

Leaders are the ones who reward performance and, given their status as role models, we try to copy their performance. Therefore, it should be easier for us to obtain rewards. Having this duality reduces the amount of deviation from the plan. In general, the culture will follow its role models faster than it follows its leadership, causing possible misalignment. However, when the role models and the leaders are one, the potential for this deviation is eliminated.

The second aspect addresses when our leaders are not our role models. Problems arise when these two positions are not aligned. In this case, the leaders set the direction for the business and reward performance for attaining the business goals that they have set. These rewards are most often in the form of job security and money. The other things that they can use as rewards are promotion and job content, both of which fit the esteem level of the Hierarchy of Needs. However, these rewards for change are undermined by the rewards provided by the culture and any role models who are not aligned with the leaders. The role models' rewards fit the level of non-romantic love within the hierarchy. Remember that before you can be concerned about esteem-level rewards, you need to satisfy those at the non-romantic love level. These are provided by the culture and the role models, not by the leadership.

People in the organization can not simply abandon their leaders and the direction they set. If they do, they may find themselves without employment. Yet they also can not abandon the rewards provided by the role models and the culture in which they have membership. What are people to do? The most prevalent answer is that they walk a tight rope between the two. They comply with the new business direction set by the leadership, but most often only enough to get by. Their real allegiance is to their role models and the culture. Their hope is that the current leadership will move on. They are then replaced with new leaders and a new program; the cycle begins again.

Those embedded within the culture clearly recognize that most often the role models, the rewards they provide, and the culture will last while the leaders are "here today and gone tomorrow." We have all experienced this when a new leader and a new program are introduced. Do we work to seek the rewards that this program du jour offers? Or do we stay with the rewards anchored in the culture and promoted by the role models? The answer is obvious – we give lip service to the leader-sponsored change and stay with the culture.

Yet waiting for leaders to leave doesn't help the organization improve. There is a problem in the model in which the leaders are promoting change and the culture is promoting status quo. Suppose that you are a leader promoting a reliability approach to the work. However the organization is reactive and led by individuals (everyone's role model) who are extremely reactive in their approach. The once-per-year salary increase and promotion to reliability-related positions that the organization views as inferior will not be a sufficient reward to drive change. The rewards provided by the reactive role model are far superior and very closely tied to the work of the organization.

Therefore, what can you as a leader do using rewards to promote change? The solution lies in the area of negative rewards. To accomplish the necessary changes in this example, you need to remove the role models from their position of power. At the same time, you must make cultural alignment with this individual unattractive to the organization.

I saw this approach achieve remarkable results. In a prior job, I worked for an individual who was reactive, who was the role model for the organization, and who was clearly not aligned with the maintenance leadership. The problem was that our equipment reliability was poor and reactive maintenance was not the answer. The manager knew he needed to make a change and refocus the organization on equipment reliability. To accomplish this, he promoted the role model to a new assignment called "special projects" outside of and away from the day-to-day work. He then filled the vacated position with a manager who was aligned with his reliability vision.

The interesting aspect of this plan was that the special projects position had no projects. The role model was essentially in a position without work. It was immediately clear to the organization that the former manager was being punished for his reactive approach and that their failure to embrace the proactive way of working could result in the same reward. This plan had an impact on the organization because years later people still joked that the last thing you wanted was for your manager to assign you to a special project.

Rites and Rituals

We know that rituals are the things that we do as part of our day-to-day work. If we are promoting a reliability-focused change, then these things we do will be altered. Rewards reinforce performance, showing themselves as rites performed by the organization. For example, in a reactive work environment, the reward for a quick response may be praise from production at the morning meeting. To promote change, we need to change the rituals as part of the work process redesign. However, if we are to use rewards to support change, a more important aspect than changing the process is changing the rites and the rewards for performance. We need to establish new rites to reward the new behaviors.

In the reactive mode, things break and maintenance rushes in to make the repair (ritual). At the morning production meeting, those involved receive praise for saving the day (rite). To make a successful switch to a proactive model, these rituals and their supporting rites must be changed. A reliability-based work environment will have fewer failures requiring less emergency response (remove the ritual). However, when emergencies occur, the organization needs to recognize, but not praise the response (change the rite). Instead, they need to treat the failure as a negative event (change the ritual) and use it to drive a failure analysis process (a new ritual). Once the analysis has been completed and corrective action taken to prevent future failure, praise should be given to those who solved the problem (new and redirected rite).

This is only one example of how rituals, their supporting rites, and the rewards can be altered to promote a new way of working. In your work processes, you need to do the analysis and make these changes. Old rituals and their rites, along with rewards for the wrong things, can certainly hamper your chances for success.

Cultural Infrastructure

Storytellers, gossips, whisperers, and keepers of the faith all have great pride regarding their sacred positions within the cultural infrastructure. The reward they receive is their feeling of importance as they carry out their cultural duties. As we have discussed in other chapters, you need their support in order to implement successful change. However, you need them to provide, in a negative context, information about how things were done, thereby dissuading the organization from continuing. You also need them to provide the "new way things are done around here."

The reward that you provide them is allowing them to remain in their cultural infrastructure roles. As a manager, moving people out of their current roles through reassignment or, if necessary, more drastic means is within your power. Yet this is not really what is needed. You need these people as allies, not as disgruntled employees in someone else's department.

To accomplish this task you need to identify the members of the cultural infrastructure, have them recognize the reward you are providing, and then establish them as allies in your change effort.

Next you need to provide them with ammunition. They need information first about the positives of change and second about the negatives of not changing. In a sense, they are going to be your sales force in this area. Using the cultural infrastructure, they can reach people in the organization far more easily than you can as manager. You also need to give them something to promote throughout the cultural infrastructure underground. Visible evidence of a new way of working and the potential rewards for success should provide this material.

I once worked on a project whose success was very important for the success of the business. To accomplish the project, we divided the work into smaller units, handling the pieces with sub-teams. It was suggested that these teams be composed of a cross section of the entire organization, including members of the workforce. Someone suggested that we select some of the best hourly workers to fill these roles. We didn't do that. Instead we identified the key members of the cultural infrastructure, then placed them on the sub-teams. We recognized that once they were convinced of the importance of success, they would be a major force within the workforce to promote the idea. It worked.

16.8 Onward to the Web of Cultural Change

So far in this book, we have discussed the four elements of culture and the eight elements of change. We have also addressed how these two sets of change elements relate with one another. In the following chapters, we take this one step further with the introduction of the Web of Cultural Change in Chapter 17 and how to analyze the web's information in Chapter 18. This web diagram, the accompanying survey, and the process of analysis should provide you with great insight into your cultural strengths as well as area where you can improve.

The Web of Cultural Change

17.1 Introduction to the Web

The purpose of this chapter is to introduce the Web of Cultural Change. In my prior book *Successfully Managing Change in Organizations: A Users Guide,* I created an eight-spoke web model that readers can use to determine change readiness within their company. It is essentially a radar diagram with eight spokes. Each of these spokes is one of the eight elements of change. In the appendix of *Successfully Managing Change in Organizations: A Users Guide* and on a disk that comes with the book, there is a survey of four questions for each of the eight elements. Each question, when answered, adds from 0 to 5 points to the score for that element. Therefore, for each element, you can conceivably obtain a score of 20 points, assuming that your firm receives the maximum score for each question. A completed Web of Change diagram for a typical company looks like Figure 17-1.

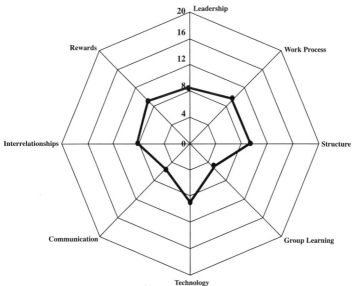

Figure 17-1 A Web of Change diagram

A score of 20 for any element places it on the outer ring. This is the best score possible for any element. Lesser scores are relegated to the inner rings with the lowest score being a 0, placing that element's score in the center. The value of this effort is to show the readiness of your company for change.

Two important aspects are associated with the scoring. First, you want to see high scores for the elements individually. High scores indicate that the things that need to be done to prepare for change are in place. Second, you do not want to see one element score much higher or much lower than the others. Otherwise, the elements are out of alignment – either positively or negatively. It is far preferable that all of the elements are relatively equal. In this way, you can be assured that your progress is being made uniformly across all of the eight elements of change.

Another benefit to the web is that elements with low scores can be analyzed in detail, then actions can be implemented to correct the deficiency. This is not to say that the web is statistically accurate – it is not. However, it certainly is an indicator whether something is not right in your company's world of change management. In addition, by reviewing the questions and their respective scores, you can perform an analysis to find out why.

The web model in this book is somewhat different. In the Web of Cultural Change, we are not seeking to understand readiness for change. Presumably you are already involved or planning to soon be involved in a change effort within you company. My purpose with the Web of Cultural Change is to assess each of the eight elements of change relative to the four elements of culture that we have been discussing throughout this text. If you fail to change the organization's culture, then over time any change initiative you put in place will fail. The Web of Cultural Change will help you identify the areas of weakness with cultural change just as the Web of Change helps you identify areas where your change readiness needs improvement.

How the Web of Cultural Change Was Built

When creating the Web of Cultural Change, I was confronted with a multi-faceted problem. First, I wanted to show all of the relationships on a radar diagram. Doing this allows the users to see the connections among the eight elements. Second, I wanted to show the relationships of the eight elements of change to each of the four elements of culture

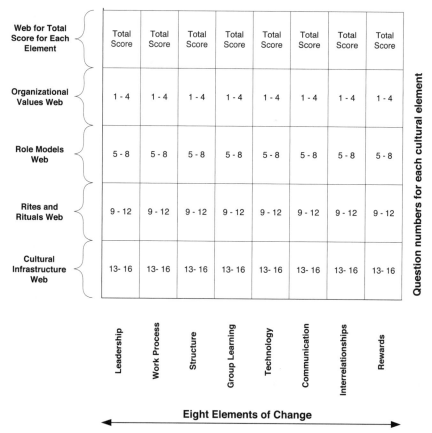

Figure 17-2   Web of Cultural Change – question set up

separately. Although I saw value in a single web diagram — so that you could see each of the eight element's scores in a composite — I also wanted to show the relationship of each of the four elements of culture separately compared to the eight elements on the diagram's spokes.

I accomplished this multi-faceted goal by creating sixteen questions for each of the eight elements of change. These sixteen questions are divided up into four sets of questions, one set for each of the four elements of culture. The first four questions relate to organizational values, the second set to role models, the next set to rites and rituals, and the last set of four questions to the cultural infrastructure. In this manner, by looking at any group of four questions across the eight elements, we can create a web diagram for just that element of culture related to the eight elements of change. Figure 17-2 shows the way that the web of cultural change questions were developed.

With the questions set up in this fashion, we can obtain either an overall web diagram or individual web diagrams for each element of culture as it relates to the eight elements of change. These web diagrams will provide a multi-faceted view. The analysis of these views will be discussed in Chapter 18.

Rather than have everyone create their own webs, I have included a CD with this text. The CD has a number of beneficial tools in addition to the web model. The key item on the CD is an Excel spreadsheet that includes a separate tab for each of the eight elements of change, including all of that element's questions and a place for you to enter your scores for each.

As you go through this process of answering the questions, a separate tab within the spreadsheet is creating your overall web of cultural change diagram as well as a separate diagram for each of the four elements of culture. This tab is labeled Baseline because it is your starting point for the process. There is another tab that is labeled Reassessment; this tab will be used for future survey scores. By having both a baseline and a set of scores after you have had a chance to do some cultural change work, you will be able to see where you improved as well as those continuing or new areas for improvement.

Within the baseline tab, there is a good deal of information. All of the scores from the individual tabs are summarized and an average is created for use in several of the charts. There is also a section where the groups of scores related to the four elements of culture are summarized so that these individual charts can be created as well. This tab has been password protected so that it can not be altered. Each of the sub tabs has also been password protected in order to preserve the web model and allow it to work as designed.

17.3 The Charts

The charts are also represented in both the baseline and the reassessment tabs. The baseline shows your original scores. The reassessment shows both the original scores and the scores that you obtain when you retake the survey after undertaking some cultural change corrective action.

The baseline tab has the following eight charts:

Cultural Web of Change

Figure 17-3 shows a representative chart in which the total score for each element is provided. These scores are based on the answers to the sixteen questions provided for that element. Because each question can receive a maximum of 5 points, the maximum score for any element is 80. If every answer is scored as "strongly disagree," then that element will have the lowest score of 16. I have made it extremely difficult, if not impossible, to have a very poor score – the inner circle on the web. My reasoning is that I do not believe that any firm is that poor. Yet any scores that received a strongly disagree score of 1 will stand out for future analysis.

Actual Score vs Average Scores

Figure 17-4 shows how the scores for each element compare with the average total score. The process of cultural change needs to proceed uniformly; any score with significant deviation from the average is either changing too fast or too slow when compared to the others.

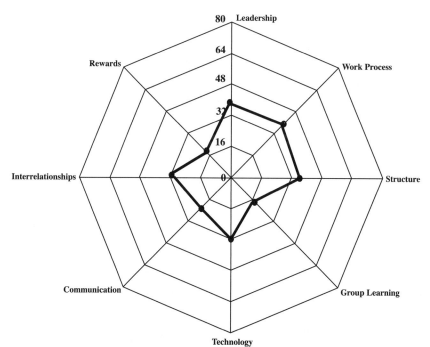

Figure 17-3   The baseline Web of Cultural Change

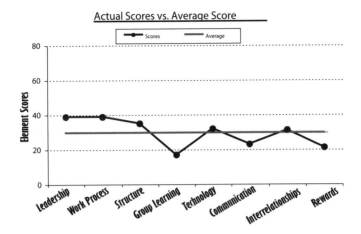

Figure 17-4   Actual scores vs. the average

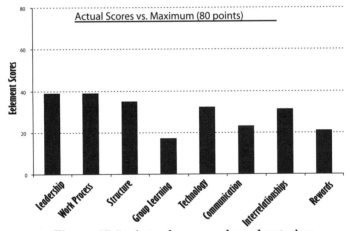

Figure 17-5   Actual scores – bar chart view

Actual Score vs Maximum (80 Points)

This chart gives the users a different look at the data. It simply represents the scores as bars with the maximum being the score of 80 points per element.

Actual Scores – Deviation from Average

This chart provides another way to look at how the scores for each of the elements deviate from the average. In this case, the chart shows both plus and minus deviation because it uses the average score as the baseline. This chart will give you a clear indication if any of the elements is out of alignment with the others.

Additional Charts

The other four charts are individual web diagrams for each of the four elements of culture. The only difference between these four charts and Figure 17-3 is that each of these is for a separate element of the culture – organizational values, role models, rites and rituals and the cultural infrastructure.

How the Web of Cultural Change Works

The web diagram works either of two ways. You can use the CD to answer the survey questions, then allow the Excel spreadsheet to create your web. Or you can copy the survey questions from Appendix 1, along with a blank web diagram provided in Appendix 2, and make your own. The former is obviously easier, but for those who want to go through the process the latter is offered.

The web model works very simply. You answer the questions by scoring from 5 points for strongly agree with the statement down to 1 point for strongly disagree. Answer all of the questions so that the score is not biased when the element's values are added and portrayed on the web and the other charts.

In Chapter 14 on communication, we discussed the need not only to send the message, but also to make sure that it was understood. There

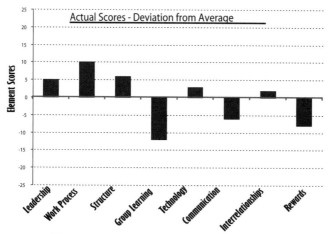

Figure 17-6   Deviation from the average

is no easy feedback mechanism with a book, yet it is my desire that there be clarity when answering the questions. Therefore, you will find in Appendix 3 a list of all of the survey questions that I thought could use some additional clarity along with my thoughts about each. In this way, if you are unclear about a question, you can look it up in the appendix to see what point I am trying to make.

17.5 An Example

In order to assure a clear understanding of the survey, how it gets assembled within the Excel spreadsheet, and what it looks like when shown as the web of cultural change, I offer the following example. In this example, I do not show the three bar or line charts that are represented on the baseline or reassessment pages, only the primary web diagram.

Suppose that your plant is going through a change process designed to convert the site to a reliability-focused work culture. The initiative seems to be having problems. Barriers to change are being encountered

Summary Baseline Data and Charts

	Max Score	Baseline Score	Baseline vs Average
Leadership	80	39	9.38
Work Process	80	39	9.38
Structure	80	35	5.38
Group Learning	80	17	−12.63
Technology	80	32	2.38
Communication	80	23	−6.63
Interrelationships	80	31	1.38
Rewards	80	21	−8.63
Average		29.63	

Figure 17-7   Example – tabular results

and the organization often finds itself unable to clear these hurdles. The site conducts a web of cultural change survey in an attempt to pinpoint the problems. The survey results are shown on Figure 17-7 and the resultant web diagram is shown in Figure 17-8.

In Figure 17-7 we see the tabular results of the cultural change survey and in Figure 17-8 the web diagram. Remember, these are based on the total scores of the sixteen questions from each section. The questions are divided into sets of four so that individual web diagrams can be developed for each of the four elements of culture. In addition, the rings of the web have significance. The inner ring (0 to 16) is a very poor score, the next ring (17 to 32) is poor, the next (33 to 48) is average, the next (49 to 64) is good and the outer ring (65 to 80) is very good. From our example, we see that we have five scores within the average ring and three in the poor ring. This information and the sub web diagrams will be very important for the analysis phase in Chapter 18.

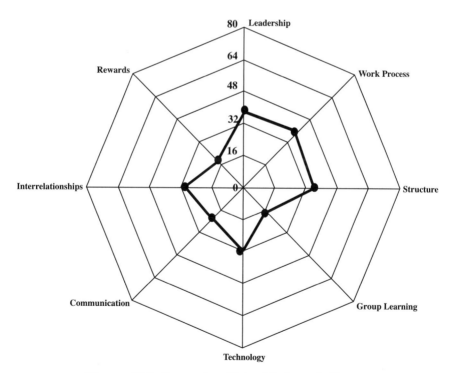

Figure 17-8 Example – Web of Cultural Change

17.6 Introduction to Reassessment

Just as it is important to get a good baseline score so that you can focus your analysis, it is also important to reassess your progress after a time period – usually six months — in order to see what progress has been made. The reassessment is handled with a separate tab within the Excel diagram. The charts provided are similar to the ones in the baseline; however, they show both the baseline and the reassessment scores. In this way, you can truly focus your attention on areas that have not improved, areas that you thought would improve but didn't, and in some cases areas that may have gotten worse as a result of actions that you have taken.

17.7 Large Group or Small Group Surveys

A question often comes up regarding how you should conduct the web of cultural change survey. Unlike *Successfully Managing Change in Organizations: A Users Guide* where the web questions were generic and could be answered without an understanding of the material in the text, this does not hold true for the web of cultural change. As you know from reading the text, there are concepts, terms, and definitions that must be clearly understood before one can take the survey and return accurate results. Therefore, the first step to approaching the web survey is to provide participants with the survey information contained in the text, making certain that they understand the material before proceeding. The group participating can be as small as one person (you) or as many as you wish. Certainly a broader population of people answering the questions will provide you with a better web diagram, but getting everyone to allocate the time to understand the material prior to the web survey may prove somewhat difficult.

In order to support your effort in this area, the CD that comes with this book also contains a Power Point presentation, with voice giving you an oversight into the content of the text. Although it does not provide all of the information contained in this book, it will give you a mechanism to educate your team and position them to conduct the survey.

There is also a spin-off benefit if you and your team are able to allocate the time required. When people in the future speak about the need to change the organizational culture, those who have participated in the learning process will understand what this truly means and how difficult the effort will be for the organization.

Assessment and Corrective Action

Chapter 18

18.1 Assessment

In Chapter 17, I discussed the Web of Cultural Change, the accompanying survey, and the charts the survey produces. The survey provides both a baseline and a set of reassessment charts and score tables. Thus, you can clearly see where you started (baseline) where your progress has taken you over time (reassessment). Each of these assessment tools has value.

Baseline

These charts show you a glimpse into the way things are at present. The areas of most concern are those elements that score in the poor (17 to 32 points) and very poor (0-16 points) range; these are the areas where you are having the biggest problems. Although each element tells a story independently, they are also closely interrelated. Therefore, an average score in one of the elements may be the result of a poor scores in another.

Another important aspect of the baseline information is to see if one or more of the elements has a large deviation from the average – either negatively or positively. In either case, this information tells you that one of your elements is out of synchronization with the others. Due to the interrelationship between all of the elements, large deviations from the average should not occur. Determining the reason for them is part of the analysis.

Reassessment

These charts show you how things have or have not improved over time. The reassessment process enables you to identify your issues in the baseline analysis, develop corrective action plans, implement the plans, and then, at some time in the future (usually about six months), to reassess yourself and see what has happened. In some cases, things will get better. Others may get worse and a few may remain neutral without changing. If you are truly working to improve, knowing this information is extremely important because it will enable you to refocus your efforts – possibly in a different direction based on the new information.

The Web of Cultural Change has an additional benefit that applies in both the baseline and reassessment modes. Because each of the elements has sixteen questions and the questions have each been subdivided into blocks of four for each of the four elements of culture, you can fine tune your analysis and examine the effects of one or more of the four elements of culture on the eight elements of change.

For example, a low score in the Leadership element is composed of the sum of the scores from all sixteen questions. Upon a more detailed examination, you may determine that the lowest set of scores is coming from questions one through four, which focus on organizational values. This knowledge will enable you to focus your attention on this element of culture in order to try to improve the leadership element. This drill-down capability provided by the Web of Cultural Change further enhances your analysis and potential improvement capability. A detailed example of this will be shown in Section 18-4.

In order to improve, you need to conduct the assessments and do the work necessary to carry out a complete analysis. Once completed, you still need to put your findings into action and closely monitor the process. Remember: there is no magic pill which corrects all of the issues. You and your firm will need hard work and commitment to see the entire process successfully through to the end.

18.2 The Technique for Analysis

There are many ways to analyze the information that is provided by the Web of Cultural Change. The one that I will describe seems to be the best and most complete method for attacking a very complex issue – changing organizational culture.

The majority of those in the reliability or maintenance business have heard of or utilized the various forms of root cause failure analysis, or RCFA for short. In this model, you identify the problem. Then by asking why the problem exists, you drill down through several layers to reach the root of your problem. Once you get to this point, you identify and implement solutions to solve the root cause. Taking these steps ultimately helps you solve the problem. A single root cause is often an oversimplified answer; most of our issues have more than one root. However, if you know the methodology associated with a simple solution, you'll find it only takes a little more work to extrapolate the process to a problem with more than one root.

In the reliability and maintenance world, RCFA is applied to determine why mechanical things fail. The purpose is to remove the failure mechanism and eliminate future failures.

Suppose that you and your organization are rolling out a work process that refocuses the plant away from the former day-to-day reactive maintenance to a new way of working. This new process involves detailed work plans, with production scheduling the work that is ready to be performed one week prior to execution. In this process, the work crews will work only on jobs that are ready to work, have been prepared in advance, and are on the schedule. The training and rollout of the process appear to go very well and everyone seems to be working in accordance with the new process.

Then the process is audited by a group of objective in-plant observers. The results are not what were expected. The process that had been rolled out is nowhere to be found. Yes, production is selecting ready to work jobs for the next week. But when the next week comes, the maintenance crews are diverted to whatever production believes is most important, even if the work is not planned and not on the weekly schedule. The audit team discovers not only that this is not a random event, but also that disregard of the new process is actually the norm.

This problem needs to be corrected if your new planning and scheduling process is going to be successful. You need to search out the root causes of the problems that your new process is encountering. Just as maintenance and reliability professional conduct root cause failure analysis (RCFA) to prevent the reoccurrence of mechanical equipment failure, we need to conduct a change - root cause failure analysis (C-RCFA) for our analogous work process failure.

Step 1 – The Question and Level 1 Answers

The C-RCFA process begins by asking the question, "Why has the planning and scheduling work process failed?" It is important to make this question rather specific so that we can frame valid answers. Several answers at level 1 are identified as follows and shown in Figure 18-1:

Production believes that whatever fails needs to be fixed right away in order to minimize any chance of production loss.

The process has never really been supported by production line management. They had no involvement in its design.

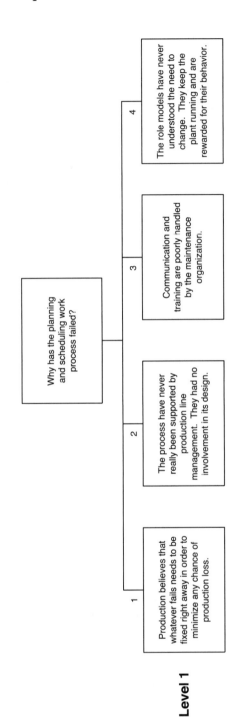

Figure 18-1 Example – Level 1 answers shown diagrammatically

Communication and training are poorly handled by the maintenance organization.

The role models have never understood the need to change. They keep the plant running and are rewarded for their behavior.

Step 2 – Level 2 Answers

Once the primary question has been answered, you need to select each answer and ask why that response has caused the planning and scheduling process to fail. For each of the level 1 answers, there may be several more answers at the next level. The analysis can start to become complicated because if we receive four answers for each of the level 1 responses, we will be faced with sixteen answers to analyze. You can imagine how cumbersome this analysis can get if we go through four levels.

However, there is a mitigating process that keeps some of this in check. When examined, many of the answers can be eliminated as possible causes, thereby making further answers for those questions unnecessary.

When we ask why for each of the answers at level 1, we arrive at the following set of level 2 answers. These are identified by the lettered items under each level 1 answer.

- Production believes that whatever fails needs to be fixed right away in order to minimize any chance of production loss.

- There are negative rewards associated with failing to meet production quotas.

- There is a fear that the spares are not good back-ups.

- If things are not made into emergencies, maintenance will never get to them.

- The process has never really been supported by production line management. They had no involvement in its design.

- They were not involved. However, their lack of commitment is addressed in item #1 above. No further analysis required here.

- Communication and training are poorly handled by the maintenance organization.

- Maintenance does not have resources for training.

- Maintenance assumes what is best for all departments.

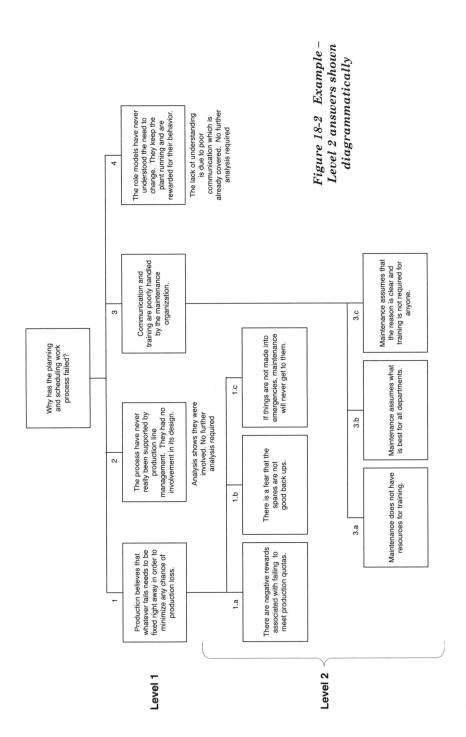

*Figure 18-2 Example –
Level 2 answers shown
diagrammatically*

- Maintenance assumes that the reason is clear and training is not required for anyone.

- The role models have never understood the need to change.

- They keep the plant running and are rewarded for their behavior.

- The lack of understanding is due to poor communication and training is covered in item #3. No further analysis is required here.

As a result of asking why a second time, as well as analyzing the validity or duplication of responses, we have identified three reasons for item #1 and three more for item #3, for a total of six responses. We also have been able to drop Items #2 and #4. This is depicted in Figure 18-2.

Step 3 – Level 3 Answers
Let us go to one further level of detail to demonstrate the process fully. Your own example may have four or even five levels to your analysis. You need to drill down to whatever level provides you with what appears to be the end of possible questions. In our example, and for simplicity's sake, we will work only on item #1 at level 3. The third level of why questions and their answers follow:

- Production believes that whatever fails needs to be fixed right away in order to minimize any chance of production loss.

- There are negative rewards associated with failing to meet production quotas.

- When discussed, there is no evidence that this is the case.

- There is a fear that the spares are not good back-ups.

- The PM program does not work and the spares are not maintained.

- The production department will not release equipment for PM.

- The PM crew is often moved to emergency work and has a large backlog of uncompleted work.

- If things are not made into emergencies, maintenance will never get to them.

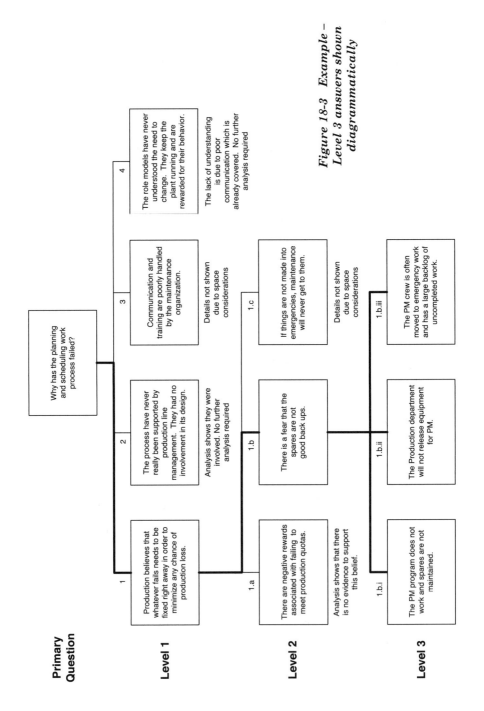

*Figure 18-3 Example –
Level 3 answers shown
diagrammatically*

- Because the planning process is not well designed.

- Conflicting production priorities.

- The C-RCFA diagram where three levels are shown in Fig18-3.

Step 4 – Analysis

Once all of the possibilities have been identified, it is time to conduct the detailed analysis of the answers that have been generated by asking why. This analysis is done at the lowest level of the answers, in our case level 3. At this lowest level, some of the answers may be eliminated. In addition, one or many will jump out as the root cause. However, it is equally possible that several answers could be the real reasons for the problem. In the case of multiple possibilities, you will need to prioritize the potential impact, then work on those where reducing the impact will have the most benefit.

In our example, we start the analysis with item #1 on level 1 and the answers we obtained at level 2 for this item. There is no evidence of dire consequences for failure to meet quota. Therefore, item 1a needs no further analysis. Furthermore, item 1c from level 2 can also be eliminated from the analysis because the evidence is that the process was well designed and based on industry best practices. Production does not need to create emergencies to get things done.

The analysis leaves us then with item 1b – there is a fear that the spares are not good back ups. Further analysis indicates that there may be something wrong with the PM program that has created a fear in production that the spare pumps are not good back-up equipment. Given this belief and the desire to maintain production, it is understandable why production does not want to wait for the planning and scheduling process when rotating equipment fails. They want it fixed now because they have no confidence in the spare equipment.

Step 5 – Fix the Root / Fix the Problem.

Our analysis seems to indicate that the PM program needs first to work and second to work correctly. If these adjustments are made, then you will have corrected a root problem.

- The PM program would be functional and the spares maintained.

- Production would release equipment for PM because they would know that the crew would not be pulled to other work.

- An effective PM program and the assurance that the maintenance crews would not be diverted to other work would increase production's confidence. They would then rely on their spares which, in turn, would reduce the need for their requirement to fix the failure now. They would now be willing to wait for the planning and scheduling process.

The entire C-RCFA diagram is graphically represented in Fig. 18-3. The pathway from level 3 corrective actions to how these actions address the problem are shown with bold connecting lines on the diagram.

18.3 How is C-RCFA Accomplished?

Change Root Cause Failure Analysis (C-RCFA) requires detailed analysis and the bringing together of various ideas about the problem and its root causes. Although it is true that the analysis could be done by a single individual, there is a risk in this approach. First, conducting the analysis alone will result in a one-dimensional set of answers. It is possible that individually you could identify the problem and work down to the roots, but it is highly unlikely. You have a built-in bias based on your background and your current position. If you were a maintenance professional in our example, there is little likelihood you would have been able to recognize the root problems from production's view point.

The second risk is that, even if you do correctly identify the optimal solution, the organization may not believe you. People resist new ideas when they have not been part of the analysis that generated the ideas, especially if they believe that they have something of value to offer. Even if you have the right answer, it may not be accepted and the problem will remain.

These types of analysis are usually conducted with a group of experts in the individual areas that appear either to affect or to be affected by the problem being analyzed. These experts provide multiple views and opinions about the problem as well as the high likelihood of buy-in and support of the corrective actions once the root cause is identified.

Another valuable approach to C-RCFA is to have the work team work with a facilitator. This is different from having someone lead the team. Leaders can be part of the effort. Facilitators are not. Their job is to remain outside the detailed analysis and help the team stay in track. They manage the meeting, the work process, the time, the interactions of the members, and the action items. Their role serves two purposes. They keep the group focused on the problem and they free up the team from having to handle the logistical issues of these events.

Many times when these efforts are conducted, the analysis shows that the root causes are people and what people have done. This is especially true when analyzing cultural change initiatives where we are dealing with role models, rites and rituals, and the cultural infrastructure. These problems will never be solved if the organization is one that seeks out the guilty and punishes them for their apparent misdeeds. Unless someone has broken the laws of the land, as discussed in Chapter 12 on Group Learning, punishment will not get you anywhere. People will not offer information needed to solve C-RCFA issues if they fear that they will be punished.

18.4 An Example Using the Web of Cultural Change

The Web of Cultural Change survey provides a great deal of information. If the analysis of this information is handled correctly, you will get root cause information about both the eight elements of change and the four elements of culture. Several rules if followed will provide you with a greater degree of success. These are:

Rules for Analysis

- You can not address everything at once. You need to select the area where things appear to be most in need of correction. In doing this, you will be addressing the biggest problem area and helping the associated problems in other areas. Remember: the elements are linked so that fixing a problem is one area has a high likelihood that you will be fixing other problems with other elements.

- The elements of change that have the poorer scores are usually driving others down as well.

- You need to conduct the analysis with a team. You need the team to bring their different views to the table for the best results.

- Take the time to do the C-RCFA analysis. It will provide you with a detailed understanding of what is going on at your plant as well as insight on how to fix the problems.

- Never assume anything. Always get hard evidence before the final solution.

- Remember that the survey is a trend, not a statistically accurate test. It gives you idea about where to look for the root problems. If someone is debating whether the score is 21 or 24, they have missed the boat. They do not understand the value the survey offers.

When you are working your way though the Web of Cultural Change and attempting to get at the details, you will need to conduct extensive interviews of the organization.

An Example of C-RCFA Analysis

Suppose that you are the reliability/maintenance manager in a plant that has implemented a proactive maintenance process. The process was designed to reduce reactivity of the workforce and to put into place reliability-focused initiatives. The change process started well with a

Summary Baseline Data and Charts			
	Max Score	**Baseline Score**	**Baseline vs Average**
Leadership	80	38	9.00
Work Process	80	30	1.00
Structure	80	33	4.00
Group Learning	80	33	−4.00
Technology	80	31	2.00
Communication	80	26	−3.00
Interrelationships	80	20	−9.00
Rewards	80	21	−8.00
Average		29.00	

Figure 18-4 Web of Cultural Change example – table of baseline data

	Organizational Values					Role Models				
	Q1	Q2	Q3	Q4	Total Values	Q5	Q6	Q7	Q8	Role Model Total
Leadership	3	3	2	2	10	3	3	2	2	10
Work Process	4	2	2	3	11	3	3	2	2	10
Structure	3	2	2	1	8	2	4	4	3	13
Group Learning	4	1	2	2	9	3	4	3	4	14
Technology	3	4	2	2	11	3	2	2	3	10
Communication	3	2	2	2	9	3	2	2	2	9
Interrelationships	2	1	1	2	6	2	2	2	1	7
Rewards	2	2	2	1	7	1	2	2	1	6

	Rites and Rituals					Cultural Infrastructure				
	Q9	Q10	Q11	Q12	Total R&R	Q13	Q14	Q15	Q16	Total CI
Leadership	2	2	2	3	9	2	2	2	3	9
Work Process	2	2	2	3	9	3	2	2	2	9
Structure	4	2	3	3	12	2	2	2	2	8
Group Learning	3	2	2	3	10	1	2	3	2	8
Technology	3	3	2	2	10	2	3	2	3	10
Communication	2	2	2	2	8	2	2	1	1	6
Interrelationships	2	1	2	2	7	2	3	2	2	9
Rewards	2	2	2	2	8	1	1	1	2	5

Figure 18-5 Example table of data by cultural element

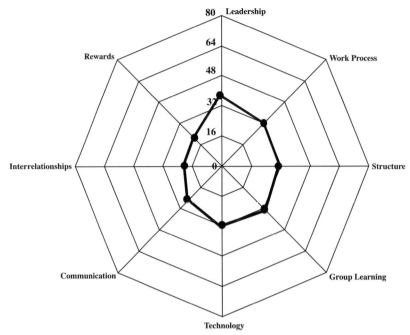

Figure 18-6   Example – Web of Cultural Change

great deal of enthusiasm and apparent buy-in. However, after six months you have been seeing more and more visible evidence that things are not well. Many of the reactive processes that you believed were eliminated appear to be present still. You desire to find out what is wrong and, more important, to put corrective actions into place to fix the problems.

To accomplish this goal, you form a cross-functional team with a team facilitator to keep the effort focused and on track. After some training in the concepts of organizational culture, your team conducts the survey across a wide population of the plant. The tabular results are shown in Figure 18-4 for the eight elements of change and in Figure 18-5 for the four elements of culture within each of the elements of change. The individual scores from this example's survey are provided in Appendix 4.

The web diagrams are displayed in figures 18-6 through 18-10.

Figure 18-6 represents the composite baseline score. Each element's score are the sum of the answers to all 16 questions in that section.

Figure 18-7 represents only the scores for organizational values. This

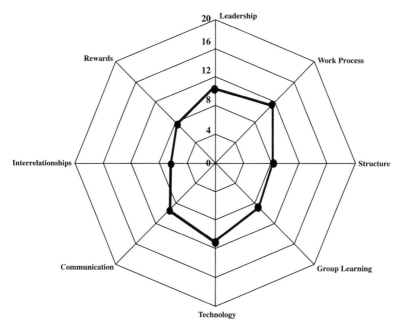

Figure 18-7 Example – Web of Cultural Change (organizational values)

data is obtained from questions 1 through 4 in each of the survey elements.

Figure 18-8 represents only the scores for role models. This data is obtained from questions 5 through 8 in each of the survey elements.

Figure 18-9 represents only the scores for rites and rituals. This data is obtained from questions 9 through 12 in each of the survey elements

Figure 18-10 represents only the scores for cultural infrastructure. This data is obtained from questions 13 through 16 in each of the survey elements

Analysis Step 1

The first step is to rank the baseline scores for the eight elements from low to high. We want to start our analysis with those that score the lowest because here is the indication of the biggest problems. Figure 18-11 shows this ranking

At this point we could single out interrelationships (20), rewards (21), or communication (26) as the elements with the worst score and

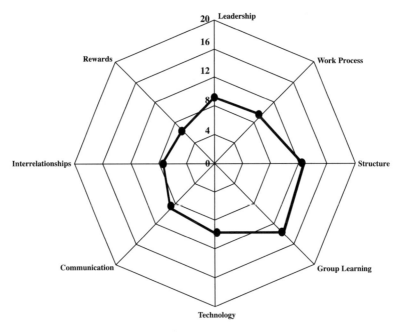

Figure 18-8   Example – Web of Cultural Change (role models)

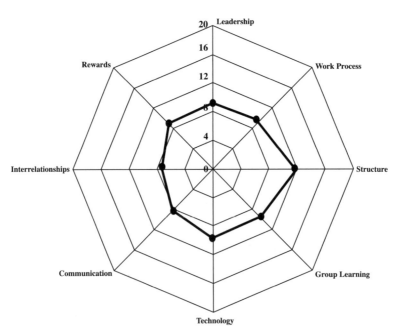

Figure 18-9   Example – Web of Cultural Change (rites and rituals)

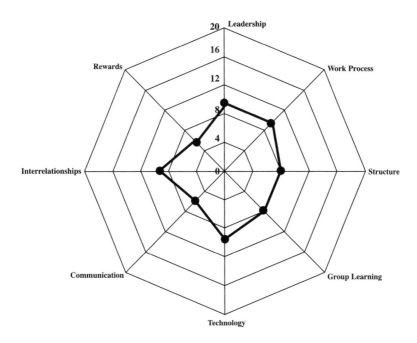

***Figure 18-10 Example – Web of Cultural Change
(cultural infrastructure)***

launch into our analysis by asking, "Why is the element of _____ so poor?" Because interrelationships has the lowest score, let us select this one for the detailed analysis. Interrelationships score for the four elements of culture are:

Organizational Values	6 points (poor range)
Role Models	7 points (poor range)
Rites and Rituals	7 points (poor range)
Cultural Infrastructure	9 points (average range)

Analysis Step 2

Now that we have this information, we need to go back to the sixteen questions in the interrelationships section, then make statements about what we have learned from the scores and the analysis to this point. Here is where the team can conduct blame-free interviews to get a firm grip on the reasons why the scores are so low. The reasons obtained in this example follow:

Organizational Values   6 points (poor range)

Organizational values may exist, but they are not shared across the plant's departments. As a result, people are not always making decisions based on the same set of values. These decisions have caused confusion and resulted in poor interrelationships among these groups.

People are not open to change. Poor interrelationships with a lack of common values support this behavior.

One set of values clearly does not apply across the organization.

Role Models   7 points (poor range)

The four scores (three questions with a score of 2 and one with a score of 1), indicate that there are poor interrelationships between the role models and that the role models are not trusted. Change can not be built on this poor foundation.

Rites and Rituals   7 points (poor range)

Horizontal and vertical interrelationships are poor. This would result from groups and role models that can't work together.

Cultural Infrastructure   9 points (average range)

Although this score is a 9 (average), it is right on the border. When examining the scores, it is apparent that good interrelationships are not promoted with those in the cultural infrastructure.

At the end of step 2, you can launch into the C-RCFA for the element of interrelationships. There is enough information to analyze and reach some root causes to the problems in this area. If you choose this direction, then you would follow it through until you reached the roots for this element. Once that route is complete, you would move on to rewards, which was the next lowest scored element.

Analysis – Step Three

Another step that can be added to your analysis may prove far more valuable in correcting cultural problems within the plant. This step takes the lowest score of the four elements of culture within the element of interrelationships – organizational values — and examines it within each of the eight elements of change. The value that results from this

Summary Baseline Data and Charts

	Max Score	Baseline Score	Baseline vs Average
Leadership	80	38	9.00
Work Process	80	30	1.00
Structure	80	33	4.00
Group Learning	80	33	−4.00
Technology	80	31	2.00
Communication	80	26	−3.00
Interrelationships	80	20	−9.00
Rewards	80	21	−8.00
Average		29.00	

Figure 18-11 Example – survey Eight Elements of Change (low to high)

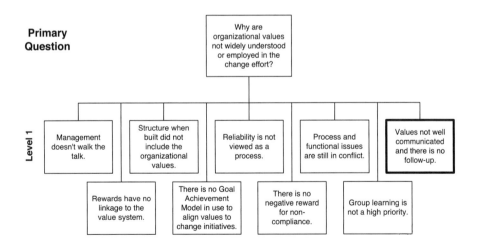

Figure 18-12   Example – C-RCFA at level 1

step is that you work on correcting a single cultural element across the entire group of the eight element of change. To continue this line of analysis, let us look at the organizational value score for each of the eight elements. They are as follows

Leadership	12 points (average)
Work Process	11 points (average)
Structure	8 points (poor)
Group learning	9 points (average)
Technology	11 points (average)
Communication	9 points (average)
Interrelationships	6 points (poor)
Rewards	7 points (poor)

Next go back and look at the questions and how they were answered. Once again, interviews with plant personnel will increase your insight. What is learned from the answers about organizational values within each of the elements of change follows:

Leadership   12 points (average)

The values exist, but there has been no follow through. The practice of the values has not been demonstrated nor have they been audited.

Work Process   11 points (average)

No auditing process

Data integrity is not part of the process

Structure   8 points (poor)

Not clear that the values are built into the structural design. It appears that all the organization did was to move the boxes.

The process vs. functional issues have not been addressed.

The structure believes that reliability is a department. This makes it difficult to promote reliability-based values across all departments.

Group Learning   9 points (average)

Type 1 learning is active, but not Type 2.

The organization is not very supportive of group learning.

There is not a significant effort being put forth to close perform-ance gaps once they have been identified.

Technology   11 points (average)

There is support for the values within the functionality of the site's technology.

The technology is not part of any improvement plan, so over time it may fail to support future processes.

Communication   9 points (average)

Values have not been well communicated. There is no process to assure understanding or receive feedback.

There is no process to keep the organization informed.

Interrelationships   6 points (poor)

Organizational values may exist, but they are not shared across the plant's departments. As a result, people are not always mak-ing decisions based on the same set of values. This has resulted in poor interrelationships among these groups.

People are not open to change. Poor interrelationships with a lack of common values would support this unwillingness to alter the status quo.

One set of values clearly does not apply across the organization.

Rewards   7 points (poor)

There does not appear to be a connection between values and the reward process.

The reward process is not understood, there are no ties to the Goal Achievement Model, and there are no negative rewards for not following the values.

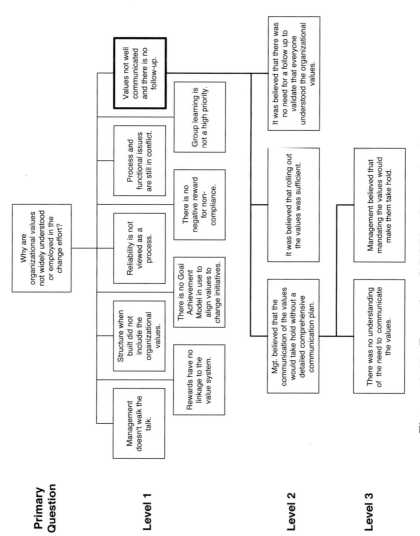

Figure 18-13 Example – C-RCFA all three levels

Analysis – Step 4

A great deal of information is provided from step 3 of the analysis process. We can now use this information to begin the C-RCFA process. Some of the details from step 3 can be eliminated because they are not of significance; some of this information could even actually apply to the second level of the C-RCFA. Figure 18-12 shows the first level of the C-RCFA for the element of organizational values across the eight elements of change.

To continue the example (on a limited scale), let's look at the level 1 reasons for our organizational value problems. Let's also focus on one of these reasons, shown with a bold outline in Figure 18-12, "Values were not well communicated and there was no follow up." Using this reason, we can go on to the next level of the C-RCFA. Through our analysis at this level, we discover that the reason the values were not well communicated were:

- Management believed that the communication of the values would take hold without a detailed and comprehensive communication plan.

- It was believed that rolling out the values was sufficient.

- It was believed that there was no need for a follow up to validate that everyone understood the organizational values.

At level 3, singling out the reason that "Management believed that the communication of the values would take hold without a detailed and comprehensive communication plan," we discover the following details:

There was no understanding of the need to communicate the values

Management believed that mandating the values would make them take hold.

The entire C-RCFA diagram for this example is shown in Fig.18-13.

With these answers at level 3, the organization can recognize that a sound communication process is required. This process would in turn convince people that values don't take root on their own (level 2) which would explain why the values were not well communicated (level 1) which would provide a sound business reason why the values were not understood and why they never were integrated into the change effort.

As a result of the C-RCFA, training would then be put into place that would be a start in correcting this problem.

This example is a very simplified one for a very complex process, but shows that using the Web of Cultural Change and the C-RCFA process, we can identify root causes of problems that exist with the eight elements of change (steps 1 and 2) or with any one of the four elements of culture (steps 3 and 4). The results of either of these levels of analysis provides the organization with corrective actions that, when completed, will solve many of the issues identified in the survey.

18.5 Reassessment

The Web of Cultural Change also has a section for reassessment. The web diagrams in this section show the baseline and the reassessment webs both on the same chart. In this manner, you can see your improvements, the areas that may have gotten worse, and those that remained neutral. The reassessment has two parts. In the first, you can examine just the reassessment data and conduct the C-RCFA. The second type of analysis is to review the changes from the baseline. For those areas that have improved, what you really need to understand is why. This information may help you in the other areas where there has been no improvement or where the scores have slipped. There are reasons for all of these changes and that is the purpose of the C-RCFA.

I am often asked how soon after the baseline survey one needs to wait before conducting a reassessment. Although there is no specific timeline, I recommend conducting the reassessment between six and twelve months. If you don't wait long enough, things will not have changed significantly to provide you with anything to analyze. On the other hand, if you wait too long, some of the things that you would wish to fix will have integrated themselves into the process and be hard to root out.

The Web of Cultural Change, the baseline, and reassessment data that it provides coupled with the C-RCFA process can go a long way to helping you to solve problems associated with cultural change. In a sense, these tools help you identify when you are getting "off track" and provide a way to get back on.

Moving Forward

19.1 Change Is Not Really a Project

The process of change is similar, yet very different, when compared to a work project. The thing that is similar at the outset of a change effort and a project is that you and your organization MUST be ready to embark. If this is not the case, the effort will fail before it ever starts. We will discuss readiness further in Section 19.2. The other similarity is the intensity of the effort required to implement the change initiative. This book and my previous book Successfully Managing Change in Organizations: A Users Guide have been focused on this topic.

The ending is certainly different. All too often when engineers and others in the maintenance and reliability arena get involved in these efforts, they treat change initiatives as if they are projects. The impact of this approach is minimal at the outset or during the work, but significant by the apparent end. At this point, we assume the work is completed; we pay less or even no further attention to the initiative on which we have spent so much time and effort. Then when we revisit the initiative in six months, we are astounded to find little or no evidence that the initiative ever existed. If we are lucky, we may still find remnants of the effort, but it will not look like it did when we stopped paying attention to it.

When you are working in the arena of change management, never assume the work is finished because it is never finished. This fact doesn't mean that you need to give change management your full-time attention, but rather you need to put into place procedures and processes that will sustain it and allow it to grow over time

We will discuss sustainability further in Section 19.3.

19.2 You Can't Start Unless You Are Ready

Change readiness is about being able to alter "how things are done around here." Conducting an effort of this sort will have an impact on the business, the work processes, and the culture of the business. Doing this is far from easy, but being totally ready before you begin is a critical success factor in any change effort. It also is a part of the foundation that will allow you and your team to sustain the initiative into the future.

The Forces Against Change

If you fail to provide a significant readiness component, you run the risk of enabling forces within the culture that will disrupt the process. These forces, as we have seen in prior discussions, come out of the cultural infrastructure as well as people's fear of change and the unknown. Examples include:

- Why change? Things are working fine using the present process.

- People don't know or understand the details. They have little confidence they can succeed.

- People are not convinced that leadership (including the role models) are in support of the initiative. This is especially the case if there have been programs of this nature (here today and gone tomorrow) in the past.

- The support in money and resources may not be apparent. Most process changes require people being detached from the regular job to work full-time on the initiative. People may question management's commitment to this approach.

- If the initiative is mandated after being designed by a select group behind closed doors, people will not feel involved and may resist.

- People will fear of losing control over the work.

- A change always destabilizes interrelationships. This will certainly be of concern because it often takes years to build them and now they are being disrupted.

- People will be concerned for their job stability and salary.

- There most likely will be a low level of compatibility with the current culture.

These fears of the unknown fit right into the processes that exist in the cultural infrastructure. As a result, the storytellers, keepers of the faith, gossips, whisperers, and spies flourish. Unfortunately they are not flourishing as we would like them to. Instead they are passing information that adds to the fear and increases resistance. Then if any one of the forces listed above actually comes true, it becomes even more powerful, ultimately causing total disruption and possible failure. I am not suggesting that the cultural infrastructure is what causes failure. However, by failing to promote readiness, those in leadership positions have set the organization up at the outset for failure. The cultural infrastructure is only behaving naturally.

A Readiness Process

Nevertheless, the majority of these problems can be avoided by developing and executing a readiness process. Then when the change takes place, the organization is ready for the change, not surprised. This process is addressed by having a clear picture of the vision you wish to achieve, communicating your vision and high level plans very early in the process before the details are set, getting people involved with the details so that they have ownership, and providing training in the new skills needed for change. Then when the actual change takes place, people already feel invested. They will work hard to make the initiative successful. Involving members of the cultural infrastructure can help bring about the level of pre-change buy-in that you want.

Let's break this process down into its component parts.

Develop a clear vision of the future.

Having a vision that paints a crystal clear picture of a not-as-yet attained future state of the business is the cornerstone of any change initiative. You need to have a clear vision of your future, not only for those who you want to commit to it, but also for yourself and the management team. This clear picture allows everyone on the team to convey the same message throughout all phases of the effort. Without a vision, confusion will reign because those who are leading the effort each will be conveying a different message. Hearing many different explanations of what the vision looks like leads to this confusion.

Communicate the vision.

Once the vision is set, it needs to be communicated throughout the organization. This communication should be handled by the management team so that everyone can see their initial level of commitment. Additionally, everyone will hear the same message, further convincing them that all parties are aligned.

However, the vision alone is insufficient. The organization most likely believes that they are doing fine with the current set of work processes, structure, and other aspects of how the business is operated. If we really want them to embrace a new vision, they need to understand why. The "why" factor is critical! Even if people understand the vision and the next steps, but are not convinced that change is needed, obtaining their buy-in will be challenging.

Involvement

Because the detailed plans are not yet determined, a great deal of organizational readiness can be promoted by getting the involvement of people within the organization who have line-level experience. If they understand the vision and believe they are involved with developing the details of the change, the value they can add to the effort is significant. This involvement also applies to those who have key positions within the cultural infrastructure and how they interact with the process. Their involvement and buy-in will save a great deal of work later on.

Training

When we think of training associated with work process change, we invariably think about training in the work process as it is rolled out. After all, the process is new. People will need to work successfully within it. However, that type of training is not the focus here.

Instead, the training that the organization needs to support the vision covers the soft skills discussed in earlier chapters. These skills include leadership, work process concepts, structure, group learning, technology, communication, interrelationships, and rewards — the eight elements of change. This type of training provides the organization with the basic skills needed to make the actual change process successful. It is not difficult to present and, in fact, I have already developed materials for this purpose. Other training companies also have courses that address these elements.

Two things are very important if you follow the recommendation to conduct "soft skill" training. First, you must train everyone. Having only part of the organization trained will create significant confusion when decisions need to be made and action taken in the normal course of business. Those who are not trained will not understand what is being done.

Second, management itself must follow what has been taught to the letter. This is not a one-time requirement, but one that must be followed all of the time. If not, the organization will immediately draw the conclusion that management is not serious. As a result, they will not take the effort serious. Part of this requirement comes from the sense that when new values and rules of behavior are set, everyone must be held accountable.

19.3 You Can't Progress Without Sustainability

After readiness, sustainability is the second important requirement for long-lasting change. Without a sustainability process, the new work culture that is being implemented will most certainly fail. Leaders get promoted, new leaders arrive with different ideas about how work is accomplished, people move between sites, people retire or leave the firm, new hires continually arrive who have no knowledge of the culture, and other changes take place continuously. Without a sustainability plan, these events have the potential to unravel the process.

Sustainability is the ability to integrate the process changes into the organization's culture so that it will not be easily changed due to changes in the organization. The change, once implemented, needs to be viewed not as additional work, but rather how we work!

Suppose your process is built around centralized maintenance. The planners reside in the production work areas; they work with production to plan the work and assemble the coming week's schedule. The foremen on the other hand are centralized and dispatched to perform work in any area based on the work schedule. In this manner, the organization can allocate their resources to the areas with the greatest need. This process has been in place for several years and the organization's level of equipment reliability has steadily improved. The maintenance manager retires and is replaced by someone from another plant where decentralized maintenance is the norm.

If there has not been a sustainability process in place, what typically happens is that the new manager attempts to alter the work process to one that is decentralized. Often by the combination of the new manager's force and the fact that the existing process is not ingrained in the culture — it is just how we do our jobs — the organization will shift to one that is decentralized. This is very disruptive to the organization.

Many of the current initiatives and value that they deliver will be lost. When this type of change occurs, the progress of the organization can be set back months or even years.

With a sustainability plan, this constant recycle can be avoided. In fact, the site most likely would not even consider hiring someone who has even the remotest intention of changing what was in place. Why?

> The entire management team would be bought into the existing process and would have no interest in altering what was working and working well.

> The work process would be fine tuned through continuous assessment of its health and corrective action to keep it healthy.

> The structure would be aligned with the process and those involved would be working within it to achieve success. Furthermore, the interrelations established by the culture and supported by the cultural infrastructure would be very strong.

> Group learning would be a key part of the effort and would support continuous improvement in the other areas of the process.

> The technology would be in place to support the current process.

> The site would have a reward structure that supported the existing process.

In other words the existing process would be so ingrained in the site's work culture that one person, even the manager, would not be able to change it.

19.4 Sustainability Tools

Sustainability is not as hard to promote as many believe. Specific actions can be taken to help the organization and its culture accept the new process and integrate it.

Communicate the vision, work status, and other things of importance.
At the heart of any sustainability plan is communication. People need to know what is going on, especially if the changes are going to have a direct effect on them. As we have seen, accurate and timely communication also minimizes the potential of negative actions by the cul-

tural infrastructure. Therefore, communication needs to be ongoing as the change initiative is developed, rolled out, and eventually becomes part of how the work is done. There is never a shortage of things for you to communicate, so there is never a reason to miss the opportunity. Remember: the cultural infrastructure is always waiting to get into the act. If you wait for this group to handle your communication process, you may find that they may not communicate what you want communicated.

Use the Goal Achievement Model

The Goal Achievement Model is another important part of sustainability. Through this model we set in place the vision, goals, initiatives, activities, and measures that will lead to our success. Although the vision seldom changes, the other components do. This model takes these changes into consideration, providing a tool to link everything together. The model is also an excellent and easy process to understand; it makes a great communication tool.

Use the Roadmap of Change

There is little worse that a good initiative gone bad. It is hard to integrate an initiative into the organization if parts of it have a negative impact on other groups. This misalignment causes problems that the Roadmap of Change addresses. This subject is discussed in detail in *Successfully Managing Change in Organizations; A Users Guide.* Eliminating misalignment problems caused by conflicting goals and initiatives will allow for sustainability.

Specific roles and responsibilities

Everyone needs to know exactly what they are required to do as part of their jobs. The ability to sustain a process is always compromised if work is not getting done because it is not clearly assigned nor understood. Another problem area is if work is being completed by two or more people, all of whom think it is their own job. The discussion of RACI charts in chapter 10 addresses this issue.

Use of the Web of Change and Web of Cultural Change

The Web of Change as described and the Web of Cultural Change are both very important sustainability tools. The former gives you an idea of the organization's readiness for change, the latter how the four elements of culture affect the eight elements of change. Although these charts provide insight, they are not of great value without assessing the information, then developing and following up on the action items that result. Taking corrective action promotes sustainability by showing the organization that the work is not a one-time event and that areas for improvement will be addressed into the future.

Ongoing audits with action items

One of the best ways to sustain any effort is to audit it on a periodic basis. You can either audit the entire process each time or audit specific components. The former gives you a good idea of the overall status, but takes a lot longer to complete whereas the latter gives you more detailed information about a specific part, takes less time, and requires fewer resources. Each has value. You should employ both based on your needs.

These audits are most often conducted by people who are outside the part of the process being audited. In this manner, they can provide objective opinions about what they hear in the interviews and what they see as they examine the process at work. Many audits use only the interview process, assuming that if you talk with a large enough cross section, you will get at the truth. This is not entirely accurate.

I have conducted audits that were focused only on interviews as well as those that had a visual examination of the work process in action. I can say without a doubt that you will get the majority of the information you seek from the interviews, but the actual examination of the process at work often provides insight that can not be gleaned from interviews.

The audit process can also be conducted by third party consultants. However, I recommend against this course of action. Although I have nothing against consultants, if they were not a part of the design, they may develop findings from the audit that support work for their firm as opposed to gaps with the existing process and recommendations for improvement.

Finally, the audit process requires that the audit team provide a final report with recommended action items. Simply reviewing a process and identifying gaps is worthless if corrective action is not taken to correct the identified problems. Sustainability is provided because continuous improvement is achieved. The organization can clearly see that the management team is serious about making the process work. Once again, this is an excellent communication tool.

An aligned reward system

Another important aspect of sustainability is to have a reward system that supports the process that is being put in place. A reward system of this nature reinforces the new process on a continuous basis, allowing everyone to see clearly how working and supporting the new process will be rewarded. It also provides negative reinforcement for attempting to revert to the former processes.

Training for new employees and follow up

As new employees arrive, training is critical. Many companies handle training very well, providing new employees with orientation training specific to their needs. Many do not provide this training or, even when they do, seriously mis-train in "how things are done around here." One of the most important parts of a sustainability process is to indoctrinate new employees into the work process, their roles and responsibilities within it, and the culture.

Refresher training for the organization

Not only is training required for new employees, but also refresher training is needed for those within the organization. This tool works well as a way to revitalize the organization. Often it is tied to communication of how the company is doing and shows how the change process has supported the improvements.

Continuous improvement process using group learning principles

The last component of a sustainability plan is to foster group learning – both Type 1 and Type -2. Organizations that are allowed to learn from both what they have done well and what they have done poorly will improve the process, improve themselves, and sustain the effort.

These tools should be carefully crafted into a sustainability program long before the actual change is put into place. They need to be ready because rollout, in the context of an organization's timeline, is short, but sustaining what you have put in place is a continuous process.

19.5 Recap

This text has four interlocking themes. First is the concept of vision and the use of the Goal Achievement Model to develop a process to accomplish the vision through goals, initiatives, and activities. The second is a detailed description of organizational culture in the context of

reliability. The concept of cultural change is covered by the four elements of culture – organizational values, roles models, rites and rituals,
and the cultural infrastructure. Understanding these concepts has probably been new for most readers; although we speak about changing the
culture, most of us (until now) haven't really understood the difficulty
associated with such a simple statement. The third theme provided a
detailed description of the eight elements of change – leadership, work
process, structure, group learning, technology, communication, interrelationships, and rewards. In addition to the detailed discussion about
each of these elements, we also explored how each is affected by and
affects the four elements of culture.

The last part of the book introduced the fourth theme: the web of cultural change. It provided a detailed survey and method of analysis so
that the reader could determine how the topics discussed in this book fit
together and develop an action plan to correct the problem areas.

I hope that as a result of reading this text, you have a better understanding of organizational culture and how difficult it is to change. If
that is the case, then the next time you hear people say, "to improve we
need to change the culture," you will be able to help them and your
organization make this difficult transition.

19.6 Final Thoughts

As I close, I would like to leave you with these final thoughts. You
might want to copy them and post them in a spot where you can refer to
them each time you feel as if you would like to give up.

- Cultural change is an essential part of what we do and who we
  are.

- Without change we stagnate and die – as companies and as
  individuals.

- With change, we are able to constantly renew ourselves and
  continuously improve.

- Change is not easy, but it is always the right thing to do.

- Change takes persistence, commitment, and a good sense of
  timing.

- Change requires an overriding desire to do what is right, often
  in the face of seemingly insurmountable odds.

- Your motto needs to be "never give up trying" because, once you do, all is lost.

- Then one day you will find you have reached the mountain top and successfully achieved the changes you sought to implement.

- And on the horizon I guarantee you will find another mountain because the journey of the change agent never ends.

Good luck in your efforts at changing your organizational culture and helping your organization to improve.

Steve Thomas

The Web of Cultural Change Survey

A. Leadership

1. Leadership has established a reliability-focused value system.
   - A. Strongly agree   5
   - B. Agree   4
   - C. Neutral   3
   - D. Disagree   2
   - E. Strongly disagree   1

2. Leadership has communicated this value system throughout the plant.
   - A. Strongly agree   5
   - B. Agree   4
   - C. Neutral   3
   - D. Disagree   2
   - E. Strongly disagree   1

3. The value system at the site has been audited to assure that it is understood at all levels of the organization.
   - A. Strongly agree   5
   - B. Agree   4
   - C. Neutral   3
   - D. Disagree   2
   - E. Strongly disagree   1

4. The site leadership demonstrates through their behavior and actions that they believe and are following the established values.
   - A. Strongly agree   5
   - B. Agree   4
   - C. Neutral   3
   - D. Disagree   2
   - E. Strongly disagree   1

5. Role models within the organization are recognized by those in leadership positions.
   A. Strongly agree   5
   B. Agree   4
   C. Neutral   3
   D. Disagree   2
   E. Strongly disagree   1

6. The site role models support the change and new ways for conducting the work.
   A. Strongly agree   5
   B. Agrcc   4
   C. Neutral   3
   D. Disagree   2
   E. Strongly disagree   1

7. Leadership clearly understands the impact that role models have on the work process, specifically a process that is undergoing change.
   A. Strongly agree   5
   B. Agree   4
   C. Neutral   3
   D. Disagree   2
   E. Strongly disagree   1

8. Role models who do not support the change effort are reassigned or removed from their area of influence.
   A. Strongly agree   5
   B. Agree   4
   C. Neutral   3
   D. Disagree   2
   E. Strongly disagree   1

9. The site's rituals and their supporting rites are clearly understood.
   A. Strongly agree   5
   B. Agree   4
   C. Neutral   3
   D. Disagree   2
   E. Strongly disagree   1

10. The rites and rituals that are out of sync with the new process are being altered.

A. Strongly agree   5
B. Agree   4
C. Neutral   3
D. Disagree   2
E. Strongly disagree   1

11. Old rites and rituals are not being reinforced.
    A. Strongly agree   5
    B. Agree   4
    C. Neutral   3
    D. Disagree   2
    E. Strongly disagree   1

12. New rites and rituals are being reinforced by the site leadership and others.
    A. Strongly agree   5
    B. Agree   4
    C. Neutral   3
    D. Disagree   2
    E. Strongly disagree   1

13. Leadership understands the importance of addressing the cultural infrastructure as part of any change effort.
    A. Strongly agree   5
    B. Agree   4
    C. Neutral   3
    D. Disagree   2
    E. Strongly disagree   1

14. The Keepers of the Faith have been involved with the development of the change initiative.
    A. Strongly agree   5
    B. Agree   4
    C. Neutral   3
    D. Disagree   2
    E. Strongly disagree   1

15. Gossips, whisperers, and spies are identified and have been part of the change process to some extent.
    A. Strongly agree   5
    B. Agree   4

C. Neutral   3
D. Disagree   2
E. Strongly disagree 1

16. Language and symbols have been developed to support the change effort.
    A. Strongly agree   5
    B. Agree   4
    C. Neutral   3
    D. Disagree   2
    E. Strongly disagree 1

B. Work Process

1.The work process has been designed to support the organizational values.

    A. Strongly agree   5
    B. Agree   4
    C. Neutral   3
    D. Disagree   2
    E. Strongly disagree 1

2. The work process is periodically audited to assure alignment, detect issues, and address corrective action.
    A. Strongly agree   5
    B. Agree   4
    C. Neutral   3
    D. Disagree   2
    E. Strongly disagree 1

3. The work process includes the ability to accurately capture information about the plant assets in such a way that the information can be utilized in support of the process and related values.
    A. Strongly agree   5
    B. Agree   4
    C. Neutral   3
    D. Disagree   2
    E. Strongly disagree 1

4. The work process has clearly delineated how the work flow and information flow models align with each other.
   A. Strongly agree 5
   B. Agree 4
   C. Neutral 3
   D. Disagree 2
   E. Strongly disagree 1

5. The role models have actively supported the work process design efforts that focus the organization on improved plant reliability..
   A. Strongly agree 5
   B. Agree 4
   C. Neutral 3
   D. Disagree 2
   E. Strongly disagree 1

6. The role models have actively supported new and improved work processes when they are rolled out and implemented within the plant.

   A. Strongly agree 5
   B. Agree 4
   C. Neutral 3
   D. Disagree 2
   E. Strongly disagree 1

7. The role model's behavior follows the work process as designed and implemented. They "walk the talk."
   A. Strongly agree 5
   B. Agree 4
   C. Neutral 3
   D. Disagree 2
   E. Strongly disagree 1

8. Role models who are not in support of the work process are identified and corrective actions are taken. These actions could be coaching or other forms.
   A. Strongly agree 5
   B. Agree 4
   C. Neutral 3
   D. Disagree 2
   E. Strongly disagree 1

9. The rituals that existed in the former work process have been altered or removed by those who support the new process.
    A. Strongly agree   5
    B. Agree   4
    C. Neutral   3
    D. Disagree   2
    E. Strongly disagree   1

10. The supporting rites (for the rituals referenced in item #9) have been altered or removed to support the new process.
    A. Strongly agree   5
    B. Agree   4
    C. Neutral   3
    D. Disagree   2
    E. Strongly disagree   1

11. Field review of the process has shown that the former rites and rituals have been removed from the process.
    A. Strongly agree   5
    B. Agree   4
    C. Neutral   3
    D. Disagree   2
    E. Strongly disagree   1

12. When a rite or ritual that supports the old process (you can't always identify them all at the outset) is identified, corrective action is taken to alter or remove them.
    A. Strongly agree   5
    B. Agree   4
    C. Neutral   3
    D. Disagree   2
    E. Strongly disagree   1

13. The cultural infrastructure's language and symbols have evolved to incorporate the new process.
    A. Strongly agree   5
    B. Agree   4
    C. Neutral   3
    D. Disagree   2
    E. Strongly disagree   1

14. The key members of the cultural infrastructure have been identified and their support elicited in developing and rolling out the new process.

    A. Strongly agree   5
    B. Agree   4
    C. Neutral   3
    D. Disagree   2
    E. Strongly disagree   1

15. A sincere effort has been made to focus stories towards success of the new process vs. how well things were done under the old. The Story Tellers have been used to support this effort.

    A. Strongly agree   5
    B. Agree   4
    C. Neutral   3
    D. Disagree   2
    E. Strongly disagree   1

16. An effective process is in place to assure that the cultural infrastructure's communication channels (used by the whisperers, gossips, and spies) are focused in support of the new process.

    A. Strongly agree   5
    B. Agree   4
    C. Neutral   3
    D. Disagree   2
    E. Strongly disagree   1

C. Structure

1. The structure has been designed to reflect the organizational values.

    A. Strongly agree   5
    B. Agree   4
    C. Neutral   3
    D. Disagree   2
    E. Strongly disagree   1

2. Reliability and maintenance are understood to be a responsibility of multiple organizations and not just a business function. This shared responsibility is clearly reflected in the values.

A. Strongly agree   5
B. Agree   4
C. Neutral   3
D. Disagree   2
E. Strongly disagree   1

3. Line and staff work issues are adequately handled by the structure. The structure was designed to make this possible.
    A. Strongly agree   5
    B. Agree   4
    C. Neutral   3
    D. Disagree   2
    E. Strongly disagree   1

4. When the structure was redesigned for the new process, the effort involved more than simply rearranging the organizational boxes. Serious thought was put into the effort.
    A. Strongly agree   5
    B. Agree   4
    C. Neutral   3
    D. Disagree   2
    E. Strongly disagree   1

5. The structural changes were made with involvement from the leadership team and role models. It was not a case of "here is how it is going to be done,"– another mandate from management.
    A. Strongly agree   5
    B. Agree   4
    C. Neutral   3
    D. Disagree   2
    E. Strongly disagree   1

6. People in leadership positions within the structure are now or are becoming the role models for the organization.
    A. Strongly agree   5
    B. Agree   4
    C. Neutral   3
    D. Disagree   2
    E. Strongly disagree   1

7. The current role models within the structure promote the new way things are to be done.
    A. Strongly agree   5
    B. Agree   4
    C. Neutral   3
    D. Disagree   2
    E. Strongly disagree   1

8. Structural vacancies are filled with the objective of hiring individuals who will ultimately become role models and promote the new process.
    A. Strongly agree   5
    B. Agree   4
    C. Neutral   3
    D. Disagree   2
    E. Strongly disagree   1

9. The structure supports the rites and rituals as dictated by the work process.
    A. Strongly agree   5
    B. Agree   4
    C. Neutral   3
    D. Disagree   2
    E. Strongly disagree   1

10. Roles and responsibilities are clear for everyone within the structure, specifically how they support the rites and rituals associated with the process.
    A. Strongly agree   5
    B. Agree   4
    C. Neutral   3
    D. Disagree   2
    E. Strongly disagree   1

11. The former set of rites and rituals (those being replaced with the new process) have been removed from how things are done around here and audits are conducted to assure this fact.
    A. Strongly agree   5
    B. Agree   4
    C. Neutral   3
    D. Disagree   2
    E. Strongly disagree   1

12. The work locations of those in the organizational structure support the rites and rituals of the process.
    A. Strongly agree   5
    B. Agree   4
    C. Neutral   3
    D. Disagree   2
    E. Strongly disagree   1

13. The members of the cultural infrastructure whose roles have been altered by the change have been identified, and efforts have been made to smooth this transition.
    A. Strongly agree   5
    B. Agree   4
    C. Neutral   3
    D. Disagree   2
    E. Strongly disagree   1

14. A structural change plan was coupled with effective and efficient communication and used to minimize the effect of the gossips.
    A. Strongly agree   5
    B. Agree   4
    C. Neutral   3
    D. Disagree   2
    E. Strongly disagree   1

15. A structural change plan was coupled with effective and efficient communication and used to minimize the effect of the whisperers.
    A. Strongly agree   5
    B. Agree   4
    C. Neutral   3
    D. Disagree   2
    E. Strongly disagree   1

16. The Keepers of the Faith have been included in the structural change process and enlisted to support it by their cultural infrastructure involvement.
    A. Strongly agree   5
    B. Agree   4
    C. Neutral   3
    D. Disagree   2
    E. Strongly disagree   1

D. Group Learning

1. Type 1 (single loop) learning is used to support current goals.
   - A. Strongly agree — 5
   - B. Agree — 4
   - C. Neutral — 3
   - D. Disagree — 2
   - E. Strongly disagree — 1

2. The company or site is open to Type 2 (double loop) learning even though employing it may identify that its goals are incorrect.
   - A. Strongly agree — 5
   - B. Agree — 4
   - C. Neutral — 3
   - D. Disagree — 2
   - E. Strongly disagree — 1

3. Gaps identified in either of the two types of learning processes are handled by developing corrective action items and follow up to completion.
   - A. Strongly agree — 5
   - B. Agree — 4
   - C. Neutral — 3
   - D. Disagree — 2
   - E. Strongly disagree — 1

4. One of the organizational values is the need to learn. The process is in place to assure that learning is continuously taking place.
   - A. Strongly agree — 5
   - B. Agree — 4
   - C. Neutral — 3
   - D. Disagree — 2
   - E. Strongly disagree — 1

5. Role models are in place who understand the organization's need to learn.
   - A. Strongly agree — 5
   - B. Agree — 4
   - C. Neutral — 3
   - D. Disagree — 2
   - E. Strongly disagree — 1

6. The role models actively support the learning process and, where it is not working, they encourage it.
   - A. Strongly agree   5
   - B. Agree   4
   - C. Neutral   3
   - D. Disagree   2
   - E. Strongly disagree   1

7. Role models recognize the need to use the Goal Achievement Model (or similar tool) to promote learning while at the same time accomplishing the vision, goals, and initiatives.
   - A. Strongly agree   5
   - B. Agree   4
   - C. Neutral   3
   - D. Disagree   2
   - E. Strongly disagree   1

8. When new things are learned that could have a positive affect on the business of reliability and maintenance, role models are open to trying them out instead of . throwing up barriers that block new ideas.
   - A. Strongly agree   5
   - B. Agree   4
   - C. Neutral   3
   - D. Disagree   2
   - E. Strongly disagree   1

9. Rites and rituals are in place and support the learning spiral.
   - A. Strongly agree   5
   - B. Agree   4
   - C. Neutral   3
   - D. Disagree   2
   - E. Strongly disagree   1

10. Rites and rituals are focused on supporting a learning organization, not a blaming one.
    - A. Strongly agree   5
    - B. Agree   4
    - C. Neutral   3
    - D. Disagree   2
    - E. Strongly disagree   1

11. Rites and rituals have embedded within them processes to identify gaps, learn from those gaps, and take the necessary corrective action.
   A. Strongly agree   5
   B. Agree   4
   C. Neutral   3
   D. Disagree   2
   E. Strongly disagree   1

12. Rites are used as a learning and reinforcing process for the rituals. When necessary, rituals are modified based on what was learned.
   A. Strongly agree   5
   B. Agree   4
   C. Neutral   3
   D. Disagree   2
   E. Strongly disagree   1

13. The cultural infrastructure is monitored to remove learning barriers.
   A. Strongly agree   5
   B. Agree   4
   C. Neutral   3
   D. Disagree   2
   E. Strongly disagree   1

14. Active efforts are in place to promote group learning within the organization, especially within in the cultural infrastructure.
   A. Strongly agree   5
   B. Agree   4
   C. Neutral   3
   D. Disagree   2
   E. Strongly disagree   1

15. Language and symbols are being learned that represent the new process.
   A. Strongly agree   5
   B. Agree   4
   C. Neutral   3
   D. Disagree   2
   E. Strongly disagree   1

16. The Keepers of the Faith are being enlisted to mentor the learning process within the site.

    A. Strongly agree   5
    B. Agree   4
    C. Neutral   3
    D. Disagree   2
    E. Strongly disagree   1

E. Technology

1. The technology has functionality that supports the organizational values.

    A. Strongly agree   5
    B. Agree   4
    C. Neutral   3
    D. Disagree   2
    E. Strongly disagree   1

2. When decisions need to be made, the technology readily provides the needed data to the users.

    A. Strongly agree   5
    B. Agree   4
    C. Neutral   3
    D. Disagree   2
    E. Strongly disagree   1

3. There is a close tie between the continuous improvement plan for the organization and the needed technology to support it.

    A. Strongly agree   5
    B. Agree   4
    C. Neutral   3
    D. Disagree   2
    E. Strongly disagree   1

4. When the organizational values are upgraded (or changed), the technology, a critical component, is also upgraded.

    A. Strongly agree   5
    B. Agree   4
    C. Neutral   3
    D. Disagree   2
    E. Strongly disagree   1

5. When role models require information to support the work, the technology makes it available.
   - A. Strongly agree   5
   - B. Agree   4
   - C. Neutral   3
   - D. Disagree   2
   - E. Strongly disagree   1

6. The technology provides reports related to the health of the work process; they are easily obtained and can be customized to the needs of the users.
   - A. Strongly agree   5
   - B. Agree   4
   - C. Neutral   3
   - D. Disagree   2
   - E. Strongly disagree   1

7. Role models actively support technology solutions in support of business needs.
   - A. Strongly agree   5
   - B. Agree   4
   - C. Neutral   3
   - D. Disagree   2
   - E. Strongly disagree   1

8. Role models promote data integration – (a single pathway to all required information available to those who need it) instead of functional / technological information silos.
   - A. Strongly agree   5
   - B. Agree   4
   - C. Neutral   3
   - D. Disagree   2
   - E. Strongly disagree   1

9. Technology supports the rituals of the site (process).
   - A. Strongly agree   5
   - B. Agree   4
   - C. Neutral   3
   - D. Disagree   2
   - E. Strongly disagree   1

10. Technology supports the rites that in turn support the rituals.
    A. Strongly agree   5
    B. Agree   4
    C. Neutral   3
    D. Disagree   2
    E. Strongly disagree   1

11. Rituals exist within the process to permit acquisition and use of new software without having to go through large amounts of red tape.
    A. Strongly agree   5
    B. Agree   4
    C. Neutral   3
    D. Disagree   2
    E. Strongly disagree   1

12. Rituals exist to permit continuous upgrading of software products in order to stay current with the vendor's releases.
    A. Strongly agree   5
    B. Agree   4
    C. Neutral   3
    D. Disagree   2
    E. Strongly disagree   1

13. Storytellers promote technology through their stories of successful implementations and the benefits delivered.
    A. Strongly agree   5
    B. Agree   4
    C. Neutral   3
    D. Disagree   2
    E. Strongly disagree   1

14. Keepers of the Faith (mentors) promote the use of technology for sound reliability-based business decisions.
    A. Strongly agree   5
    B. Agree   4
    C. Neutral   3
    D. Disagree   2
    E. Strongly disagree   1

15. Software and hardware issues are widely communicated to diminish

the potential negatives that could be conveyed by the whisperers, gossips, and spies.
      A. Strongly agree   5
      B. Agree   4
      C. Neutral   3
      D. Disagree   2
      E. Strongly disagree   1

16. There is a good communication plan to promote new technology through language and symbols.
      A. Strongly agree   5
      B. Agree   4
      C. Neutral   3
      D. Disagree   2
      E. Strongly disagree   1

F. Communication

1. The organizational values are widely communicated.
      A. Strongly agree   5
      B. Agree   4
      C. Neutral   3
      D. Disagree   2
      E. Strongly disagree   1

2. There are on-going processes in place to assure that the organizational values are understood.
      A. Strongly agree   5
      B. Agree   4
      C. Neutral   3
      D. Disagree   2
      E. Strongly disagree   1

3. Where questions exist about the organizational values and their interpretation, feedback is provided for clarification.
      A. Strongly agree   5
      B. Agree   4
      C. Neutral   3
      D. Disagree   2
      E. Strongly disagree   1

4. There is a continuous process in place to keep the workforce informed about the organizational values and how they should be applied in the day-to-day work.
   A. Strongly agree   5
   B. Agree   4
   C. Neutral   3
   D. Disagree   2
   E. Strongly disagree   1

5. Role models are good communicators of the organizational values.
   A. Strongly agree   5
   B. Agree   4
   C. Neutral   3
   D. Disagree   2
   E. Strongly disagree   1

6. Role models are good listeners. They receive feedback that could either validate that the new process is understood or establish the need for clarification.
   A. Strongly agree   5
   B. Agree   4
   C. Neutral   3
   D. Disagree   2
   E. Strongly disagree   1

7. Role models provide continuous communication, not just one time or with minimal frequency.
   A. Strongly agree   5
   B. Agree   4
   C. Neutral   3
   D. Disagree   2
   E. Strongly disagree   1

8. Role models make themselves available to the organization to be good communicators. They view it as a key component of their work responsibilities.
   A. Strongly agree   5
   B. Agree   4
   C. Neutral   3
   D. Disagree   2
   E. Strongly disagree   1

9. Rites are clearly communicated and there is a process to check under standing.
    A. Strongly agree   5
    B. Agree   4
    C. Neutral   3
    D. Disagree   2
    E. Strongly disagree   1

10. Rituals are clearly communicated and there is a process to check understanding.
    A. Strongly agree   5
    B. Agree   4
    C. Neutral   3
    D. Disagree   2
    E. Strongly disagree   1

11. The communications regarding rites and rituals have a lot of "why we are doing this and what the benefits are" content.
    A. Strongly agree   5
    B. Agree   4
    C. Neutral   3
    D. Disagree   2
    E. Strongly disagree   1

12. Communication is widely used to reinforce the work process - the rituals, and their supporting rites.
    A. Strongly agree   5
    B. Agree   4
    C. Neutral   3
    D. Disagree   2
    E. Strongly disagree   1

13. Storytellers' communications support the process instead of how it was done around here before the change.
    A. Strongly agree   5
    B. Agree   4
    C. Neutral   3
    D. Disagree   2
    E. Strongly disagree   1

14. Keepers of the Faith (mentors) focus their communication more on the new process instead of trying to show how the old way was better.
    A. Strongly agree   5
    B. Agree   4
    C. Neutral   3
    D. Disagree   2
    E. Strongly disagree   1

15. A large amount of communication keeps everyone informed, diminishing the values of the gossips.
    A. Strongly agree   5
    B. Agree   4
    C. Neutral   3
    D. Disagree   2
    E. Strongly disagree   1

16. A large amount of communication keeps everyone informed, diminishing the values of the whisperers.
    A. Strongly agree   5
    B. Agree   4
    C. Neutral   3
    D. Disagree   2
    E. Strongly disagree   1

G. Interrelationships

1. Organizational values are widely shared by the leadership team to promote buy-in and positive interrelationships.
    A. Strongly agree   5
    B. Agree   4
    C. Neutral   3
    D. Disagree   2
    E. Strongly disagree   1

2. People at all levels work well together and are open to change.
    A. Strongly agree   5
    B. Agree   4
    C. Neutral   3
    D. Disagree   2
    E. Strongly disagree   1

3. There is a single set of organizational values that is applied across all functions. There are no functional silos where the values apply differently dependent on your function.
      A. Strongly agree   5
      B. Agree   4
      C. Neutral   3
      D. Disagree   2
      E. Strongly disagree   1

4. If audited, the site would show clear evidence that one set of values existed and was in fact being applied.
      A. Strongly agree   5
      B. Agree   4
      C. Neutral   3
      D. Disagree   2
      E. Strongly disagree   1

5. Role models build organizational trust via positive interrelationships.
      A. Strongly agree   5
      B. Agree   4
      C. Neutral   3
      D. Disagree   2
      E. Strongly disagree   1

6. Role models employ the concept of reciprocity in a positive manner.
      A. Strongly agree   5
      B. Agree   4
      C. Neutral   3
      D. Disagree   2
      E. Strongly disagree   1

7. Role models create and support their allies in an effort to support the change process.
      A. Strongly agree   5
      B. Agree   4
      C. Neutral   3
      D. Disagree   2
      E. Strongly disagree   1

8. Role models maintain both horizontal and vertical positive interrelationships.

A. Strongly agree   5
B. Agree   4
C. Neutral   3
D. Disagree   2
E. Strongly disagree   1

9. Horizontal interrelationships are a very important aspect of the site rituals.
A. Strongly agree   5
B. Agree   4
C. Neutral   3
D. Disagree   2
E. Strongly disagree   1

10. Vertical interrelationships are a very important aspect of the site rituals.
A. Strongly agree   5
B. Agree   4
C. Neutral   3
D. Disagree   2
E. Strongly disagree   1

11. Rituals that exist were built so that positive interrelationships are a result.
A. Strongly agree   5
B. Agree   4
C. Neutral   3
D. Disagree   2
E. Strongly disagree   1

12. Rites that exist (in support of rituals) were built so that positive interrelationships are promoted as a result.
A. Strongly agree   5
B. Agree   4
C. Neutral   3
D. Disagree   2
E. Strongly disagree   1

13. Storytellers promote positive interrelationships across the system by the stories that they relate.
A. Strongly agree   5

B. Agree 4
C. Neutral 3
D. Disagree 2
E. Strongly disagree 1

14. Keepers of the Faith (mentors) promote positive interrelationships as one of the best ways to get things done around here.
    A. Strongly agree 5
    B. Agree 4
    C. Neutral 3
    D. Disagree 2
    E. Strongly disagree 1

15. Role models and other members of the management team promote positive interrelationships with the whisperers to enable themselves to get the real story about what is going on with the process.
    A. Strongly agree 5
    B. Agree 4
    C. Neutral 3
    D. Disagree 2
    E. Strongly disagree 1

16. Role models and other members of the management team promote positive interrelationships with the gossips. This helps them find out what is going on with the process and, where necessary, correct miscommunication.
    A. Strongly agree 5
    B. Agree 4
    C. Neutral 3
    D. Disagree 2
    E. Strongly disagree 1

H. Rewards

1. The organizational values include a reward process that supports its success.
    A. Strongly agree 5
    B. Agree 4
    C. Neutral 3
    D. Disagree 2
    E. Strongly disagree 1

2. The organization clearly understands the reward process as well as how it is linked to the organizational values.
   - A. Strongly agree   5
   - B. Agree   4
   - C. Neutral   3
   - D. Disagree   2
   - E. Strongly disagree   1

3. The Goal Achievement Model (or a similar process) is used to promote the values through the vision–goals–initiative–activity process.
   - A. Strongly agree   5
   - B. Agree   4
   - C. Neutral   3
   - D. Disagree   2
   - E. Strongly disagree   1

4. The system provides negative or neutral rewards for courses of action that do not support the value system being utilized.
   - A. Strongly agree   5
   - B. Agree   4
   - C. Neutral   3
   - D. Disagree   2
   - E. Strongly disagree   1

5. The management team recognizes that money is not a long-term motivational reward for work well done. Alternative rewards are provided.
   - A. Strongly agree   5
   - B. Agree   4
   - C. Neutral   3
   - D. Disagree   2
   - E. Strongly disagree   1

6. The time delay between successful action and its reward is minimized wherever possible.
   - A. Strongly agree   5
   - B. Agree   4
   - C. Neutral   3
   - D. Disagree   2
   - E. Strongly disagree   1

7. Role models who promote the work process and other change efforts are visibly rewarded by the organization so that others can see what success looks like.
    A. Strongly agree   5
    B. Agree   4
    C. Neutral   3
    D. Disagree   2
    E. Strongly disagree   1

8. Role models who do not promote the change (or may even work against it) visibly receive neutral or negative rewards so that the workforce can see what failure looks like.
    A. Strongly agree   5
    B. Agree   4
    C. Neutral   3
    D. Disagree   2
    E. Strongly disagree   1

9. Rituals (where appropriate) carry timely rewards that reinforce the change effort.
    A. Strongly agree   5
    B. Agree   4
    C. Neutral   3
    D. Disagree   2
    E. Strongly disagree   1

10. Rites are designed to provide rewards that reinforce the rituals and the process.
    A. Strongly agree   5
    B. Agree   4
    C. Neutral   3
    D. Disagree   2
    E. Strongly disagree   1

11. When rewards do not reinforce the change effort, they are adjusted.
    A. Strongly agree   5
    B. Agree   4
    C. Neutral   3
    D. Disagree   2
    E. Strongly disagree   1

12. When there is a time delay that will diminish the value of a reward, an active effort is made to eliminate the delay or make it as small as possible.
    A. Strongly agree   5
    B. Agree   4
    C. Neutral   3
    D. Disagree   2
    E. Strongly disagree   1

13. Members of the cultural infrastructure recognize that one reward if they remain in roles that are key to the infrastructure is their support for the change effort.
    A. Strongly agree   5
    B. Agree   4
    C. Neutral   3
    D. Disagree   2
    E. Strongly disagree   1

14. Management actively engages and rewards cultural infrastructure members for their help in promoting the change.
    A. Strongly agree   5
    B. Agree   4
    C. Neutral   3
    D. Disagree   2
    E. Strongly disagree   1

15. Negative and / or neutral rewards are provided to those members of the cultural infrastructure who use their roles to disrupt the change process.
    A. Strongly agree   5
    B. Agree   4
    C. Neutral   3
    D. Disagree   2
    E. Strongly disagree   1

16. Spies are not rewarded for passing sensitive information. At the minimum, they are informed that they are disrupting the process.
    A. Strongly agree   5
    B. Agree   4
    C. Neutral   3
    D. Disagree   2
    E. Strongly disagree   1

A Blank Web of Cultural Change Diagram

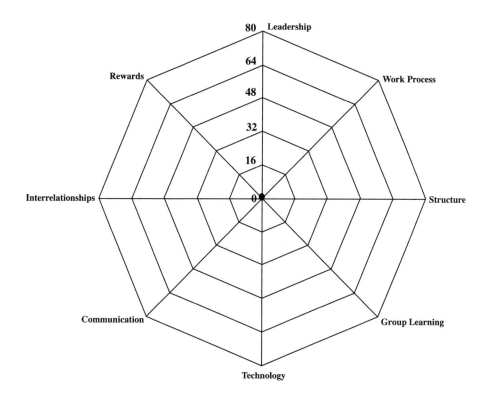

Clarification of Some of the Web Survey Questions

To promote sound communication and to make sure that the questions in the Web of Cultural Change survey are understood, I offer the following explanation of selected questions I believe may need additional clarification. The item numbers match those in the survey in Appendix 1 and on the Excel spreadsheet on the CD.

A. Leadership

1. Leadership has established a reliability-focused value system.
The value system in place is focused on change as opposed to maintenance of the status quo. For an organization that wants to promote reliability, the value system would have components in place to reduce the amount of reactivity in the work process system.

2. Leadership has communicated this value system throughout the plant.
A value system has no real usefulness if it is not communicated to those who must use it in their day-to-day activities.

3. The value system at the site has been audited to assure that it is understood at all levels of the organization.
Developing a value system and communicating it makes up only 2/3 of the process. The system must also be audited to assure that it is in place. A clearly-defined auditing process must be in place to validate its use or take the needed corrective action.

4. The site leadership demonstrates through their behavior and actions that they believe & are following the established values.
Walking the talk or demonstrating belief in the values through visible action is critical. This question tests that premise.

5. Role models within the organization are recognized by those in leadership positions.

It is important to know who the role models are because they are the people who the organization will ultimately follow.

7. Leadership clearly understands the impact that role models have on the work process, specifically a process that is undergoing change.

Whether or not the leaders themselves are the role models, they must recognize who the role models are and the importance they have in the process change.

13. Leadership understands the importance of addressing the cultural infrastructure as part of any change effort.

Typically the cultural infrastructure is not considered when change efforts are undertaken. This question determines if this omission is the case at your site.

14. The Keepers of the Faith have been involved with the development of the change initiative.

The Keepers of the Faith are mentors and what they normally mentor is the old way of doing things. In order to reverse this practice, they need to be identified and be part of the development of the new process. In this way, you can elicit their support and have them mentor the new ways.

15. Gossips, whisperers, and spies are identified and have been part of the change process to some extent.

These members of the cultural infrastructure should also be included in the work process effort. You may not want them as extensively involved as the Keepers of the Faith, but you want their support. You also don't want them passing along misinformation and slowing down the effort. Therefore, some involvement can alleviate this potential problem.

16. Language and symbols have been developed to support the change effort.

A lot can be accomplished by promoting language and symbols as part of any change effort. People latch onto these things because they want to be considered as part of what is happening instead of feeling as an outsider.

B. Work Process

2. The work process is periodically audited to assure alignment, detect issues. and address corrective action.

Implementing a work process and then assuming it has been success-fully implemented can have serious negative consequences. That is why it is very important that the process is audited, problems iden-tified and corrective action taken. These audits should be formal in nature and conducted by people who can form objective opinions.

3. The work process includes the ability to accurately capture information about the plant assets in such a way that the infor-mation can be utilized in support of the process and related val-ues.

The availability of accurate and timely data is critical to the success of any work process. This question is seeking an affirmation that it is an important part of your process.

4. The work process has clearly delineated how the work flow and information flow models align with each other.

The work process (movement of work through the system) must be closely tied to the information flow process (the movement of data and information). The work process model should have these touch points clearly identified.

8. Role models who are not in support of the work process are identified and corrective actions are taken. This could be coaching or other forms.

It is essential that people who are the role models support the work process and the other change initiatives. Those who don't and are left in positions of influence are a serious threat to long-term success. Obviously, these are good people; coaching techniques to get them to change is always the first course of action. However, failure to change needs to bring with it more serious career consequences.

11. Field review of the process has shown that the former rites and rituals have been removed from the process.

This is another audit check to see that what was going to be done was actually done in the field.

15. A sincere effort has been made to focus stories towards success of the new process vs. how well things were done under the old. The Story Tellers have been used to support this effort.

Storytellers are those people within the cultural infrastructure who reinforce how things are successfully achieved within the work process through war stories. Typically these stories reinforce the old way of working. What we want to do is to convert these people and have them tell stores that reinforce the new process.

16 An effective process is in place to assure that the cultural infrastructure's communication channels (used by the whisperers, gossips, and spies) are focused in support of the new process.

In order to reduce the potential negative impact of these members of the cultural infrastructure, we need to really open up the communication channels. This gives those who occupy these positions little to talk about, especially if what they talk about is going to be negative.

C. Structure

7. The current role models within the structure promote the new way things are to be done.

It is important that the role models support the change whether they are in formal leadership positions or not. In either case, they have tremendous influence and the change initiative needs their support.

10. Roles and responsibilities are clear for everyone within the structure, specifically how they support the rites and rituals associated with the process.

It is important that all participants in the process understand their roles and responsibilities. If they do, then the likelihood of success is far more secure than if it is not.

11. The former set of rites and rituals (those being replaced with the new process) have been removed from how things are done around here. Audits are conducted to assure this fact.

Another question regarding audits to assure that what is put in place is actually being practiced.

12. The work locations of those in the organizational structure support the rites and rituals of the process.

Part of the success of the rituals and their supporting rites is the need to have people who interact at work relatively close to one another. At times this is not possible and some alternatives need to be put into place. However, whenever possible, it is beneficial to the change effort.

14. A structural change plan was coupled with effective and efficient communication exists and used to minimize the effect of the gossips.

If you are changing the structure, you really need a detailed plan of how and, more important, why the changes are being made. Once this plan exists, it is important that it be communicated by the leadership team. Otherwise, the gossips will have a field day and the majority of what is communicated will be far from the reality of the situation. Communicate first!

15. A structural change plan was coupled with effective and efficient communication and used to minimize the effect of the whisperers.

The same rule as stated in item #14 applies here. The difference is that in this case the whisperers are passing erroneous information to the manager. Upfront communication eliminates the need for a lot of damage control.

D. Group Learning

The cultural infrastructure is monitored to remove learning barriers.

As an example, "we tried that before and it never worked" is a barrier to learning. Such barriers must be removed for progress to take place.

14. Active efforts are in place to teach group learning within the organization, especially within the cultural infrastructure.

It is important to make certain that learning is taking place, especially within the cultural infrastructure. After all, you want to promote new stories, mentoring of the new processes, and communication flow through this informal structure supportive of the change effort.

E. Technology

3. There is a close tie between the continuous improvement plan for the organization and the needed technology to support it.

All organizations have long-term improvement plans, but not all plans include a link to the organization's technology. This question tests for this linkage.

5. When role models require information to support the work, the technology makes it available.

It is essential that information be available when needed. Role models lead the organization and they often require access to timely and accurate data.

8. Role models promote data integration (a single pathway to all required information available to those who need it) instead of functional / technological information silos.

Having functional or technical information silos where only specific people within a specific function can access data is detrimental to long-term success and needs to be corrected.

9. Technology supports the rituals of the site (process).

If a work process change from reactive to proactive maintenance is to be successful, then the technology needs to support the new process – the rituals. For example, how supportive of a new planning and scheduling process would a computer system be if this portion of its functionality were seriously lacking?

10. Technology supports the rites that in turn support the rituals.

Technology needs to support the rites that in turn support the rituals if the process is going to work properly. Take, for example, the quarterly performance review – the rite that reinforces all of the rituals that take place during the quarter. To properly make the required presentation, information and presentation materials are required. These are furnished by the technology.

16. There is a good communication plan to promote new technology through language and symbols.

Most technology carries with it user language and various symbols in the form of acronyms. A good communication plan will promote these and in doing so promote acceptance.

F. Communication

11. The communications regarding rites and rituals have a lot of "why we are doing this and what the benefits are" content.

Communications always bring value. In the case of rites and rituals, especially when they are part of a new process, there needs to be a lot of explanation in the communication content. In that way, people get a chance to understand why they are doing things instead of being told without the reasoning.

G. Interrelationships

6. Role models employ the concept of reciprocity in a positive manner.

Reciprocity, as described in the text, can be a positive force in building strong interrelationships. This question tests to see if it is being employed.

9. Horizontal interrelationships are a very important aspect of the site rituals.

These include the interrelationships between peers both inside and external to your company.

10. Vertical interrelationships are a very important aspect of the site rituals.

These are the upward interrelationships with your managers and the downward interrelationships with your subordinates.

H. Reward

3. The Goal Achievement Model (or a similar process) is used to promote the values through the vision - goals - initiative - activity process.

This question tests if a process is in place to work through the vision, goals, initiatives, and activities necessary to have a successful change effort. Without a process such as this, rewards are out of the question because you will never know that you have achieved your targets.

13. Members of the cultural infrastructure recognize that one reward if they remain in roles that are key to the infrastructure is their support for the change effort.

This question assumes not only that those who make up the cultural infrastructure are known, but also that they recognize that their positions within the cultural infrastructure are in a way a reward for supporting the change.

This table represents the individual web survey scores from the example referenced in chapter 18.

	Leadership	Work Process	Structure	Group Learning
Question #	Score	Score	Score	Score
1	3	4	3	4
2	3	2	2	1
3	2	2	2	2
4	2	3	1	2
5	3	3	2	3
6	3	3	4	4
7	2	2	4	3
8	2	2	3	4
9	2	2	4	3
10	2	2	2	2
11	2	2	3	2
12	3	3	3	3
13	2	3	2	1
14	2	2	2	2
15	2	2	2	3
16	3	2	2	2
Total	38	30	33	33

Question #	Technology Score	Communication Score	Inter-relationships Score	Rewards Score
1	3	3	2	2
2	4	2	1	2
3	2	2	1	2
4	2	2	2	1
5	3	3	2	1
6	2	2	2	2
7	2	2	2	2
8	3	2	1	1
9	3	2	2	2
10	3	2	1	2
11	2	2	2	2
12	2	2	2	2
13	2	2	2	1
14	3	2	3	1
15	2	1	2	1
16	3	1	2	2
Total	31	26	20	21

Bibliography

Balmert, Paul. *Managing Safety Performance* (n.p, n.d.) (2004).

Barstow, Alan. *"Building Effective Organizations: Getting Things Done."* Masters course presented at the University of Pennsylvania (Philadelphia, PA: 1995).

Bartol, Kathryn M. and David C. Martin. Management (New York: McGraw-Hill, 1998).

Deal, Terrence E. and Allen A. Kennedy. Corporate Cultures (out of print)

(New York: Addison Wesley, 1982).

Eldred, John. "Mastering Organizational Politics and Power." Masters course presented at the University of Pennsylvania (Philadelphia, PA: 1994).

Machiavelli, Niccolo. The Prince (translated) (New York: Bantam Books, 1966).

Maslow's Hierarchy of Needs http://web.utk.edu/~gwynne/maslow.htm

Nelms, Robert. *What You Can Learn from Things that Go Wrong – A Guidebook to the Root Causes of Failure* (Virginia: Failsafe Network).

Pratkanis, Anthony and Elliot Aronson. *Age of Propaganda: The Everyday Use and Abuse of Persuasion* (New York: W. H. Freeman and Company, 1991).

Robbins, Stephen. *Organizational Behavior* (Englewood Cliffs: Prentice Hall, 1996).

Schein, Edgar H. *Organizational Culture and Leadership 2nd Edition* (San Francisco: Jossey Bass, 1992).

Schein, Edgar H. *The Corporate Culture Survival Guide* (San Francisco: Jossey Bass, 1999).

Thomas, Stephen J. *Successfully Managing Change in Organizations: A Users Guide* (New York: Industrial Press, 2001).

Figure Index

Index

X,Y,Z